D0787767

SHELL SHOCK

SHELL SHOCK

THE SECRETS AND SPIN OF AN OIL GIANT

Ian Cummins and John Beasant

MAINSTREAM PUBLISHING
EDINBURGH AND LONDON

Copyright © Ian Cummins and John Beasant, 2005
All rights reserved
The moral rights of the authors have been asserted

First published in Great Britain in 2005 by
MAINSTREAM PUBLISHING COMPANY (EDINBURGH) LTD
7 Albany Street
Edinburgh EH1 3UG

ISBN 1 84018 941 X

No part of this book may be reproduced or transmitted in any form or by any other means without permission in writing from the publisher, except by a reviewer who wishes to quote brief passages in connection with a review written for insertion in a magazine, newspaper or broadcast

Excerpts from the BBC's *File on 4* radio programme number 04VY3012LHO, broadcast on 23 March 2004, appear by kind permission of the corporation

Extracts from *The Benn Diaries* appear by kind and generous permission of the Rt. Hon. Tony Benn PC

A catalogue record for this book is available from the British Library

Typeset in Caslon, Emilyshand and Eurostile

Printed and bound in Great Britain by
William Clowes Ltd, Beccles, Suffolk

AUTHORS' NOTES AND ACKNOWLEDGEMENTS

When we were commissioned to write this book, we of course contacted Shell Centre on London's South Bank and, after four telephone calls, were equipped with the name and title of the person to whom we should formally address a letter setting out our intentions and the scope of our project.

The letter, detailing precisely what we wanted to do and what help we were seeking from Shell, was dispatched within 24 hours. What followed was the opening of a window on a weird parallel universe, a corporate *Through the Looking-Glass* world devised and delineated by John le Carré.

For our letter evoked no response at all. As days of silence turned into weeks, we began making more calls. Bizarrely, Shell initially refused either to confirm or deny that our letter had arrived. After a month had passed without a reply, we tried to speak directly to the man whose name we had been given. Shell now said – twice – that no such person was, or had ever been, employed by the company and, moreover, no such job title could be found in the corporate lexicon.

And then, two months after our first attempts at contact, a letter arrived in the second-class post from a colleague of the man who allegedly did not exist, saying Shell was not minded to help or cooperate with us and would not make available directors and senior staff for interview on currently contentious matters such as the reserves crisis.

On Shell's involvement in earlier crises of other kinds, we were directed to *A Century in Oil*, the history written for Shell – the company owns the copyright – by Stephen Howarth.

In truth, this corporately endorsed history provides a useful chronology. But anybody seriously interested in writing about a company often described as wielding more power and influence than many a small European state must acknowledge the debt owed to Robert Henrique's magisterial *Marcus Samuel: First Viscount Bearsted*, Glyn Roberts' comprehensive biography of Henri Deterding, *The Most Powerful Man in the World*, Daniel Yergin's outstanding history of the oil industry, *The Prize*, Anthony Sampson's invaluable essays on Shell in his *Anatomy of Britain* and *The Seven Sisters*, and Karl Maier's cool – and often chilling – account of Shell's involvement in Nigeria, *This House Has Fallen*.

Given that Shell declined any kind of assistance with this work, we decided that current and former employees of the company who rendered invaluable help – not least with sharp insights into a corporate culture in the throes of revolution – should remain anonymous. We take this opportunity to salute their candour, great good humour and patience.

And patience, in saintly measure, has been displayed by Ailsa Bathgate, our editor at Mainstream, whose ability to spot an infelicity at 50 paces has been given every opportunity to shine in the writing of a book not entirely without complication. Any remaining errors are, of course, entirely the responsibility of the authors.

We gratefully acknowledge all those – on both sides of the Atlantic – who provided and helped marshal an awesome volume of information. Some, in diplomatic and security services, have requested anonymity. Others, such as Beryl Angel and Mark Hollingsworth, came up with precious nuggets of raw material. Essex County Council library staff at Clacton were unfailingly courteous in helping to locate particular books and texts. Marilyn Tredree and Luke and Maria Cummins played a key role in keeping the electronic show on the road when a digitally challenged pair of old hacks created computing crises of a spectacular kind.

Very special thanks are due to Kay Cummins, who, in 40 years of cheerfully supporting an astonishing variety of obsessions, enthusiasms, whims and fancies in a dozen countries, remains the best, and best informed, sounding board in the business.

Grateful acknowledgement for sustained encouragement, professional support and, indeed, guidance on a multitude of matters is also paid to the Lady Hermione Grimston, Michael and Barbara Williams, Mary Morgan, Geoffrey Bellamy, Professor Peter Royle, Leszek Kobiernicki, Oliver Donachie, Peter Zedlewski and Raghuveer Shettigar. Tribute is also paid to Bill Campbell and Peter MacKenzie, directors of Mainstream, for their confidence in an exercise the success of which was not always immediately apparent.

CONTENTS

PROLOGUE

LIES, DAMNED LIES AND RESERVES

First sightings of the iceberg were made in London in the early hours of Friday, 9 January 2004.

For flickering on the monitor screens of bankers, brokers, oil traders and journalists was a communiqué from the energy giant Shell. In tortured language made bafflingly opaque by jargon, the company appeared to be saying that it had, as though in a fit of absence of mind, overstated its reserves by 3.9 billion barrels.

At a stroke, the British–Dutch multinational was effectively cutting its stock in trade by a fifth and at least $70 billion from future revenues.

That, when held up to the light, subjected to forensic scrutiny and decrypted by business code-breakers, was what Shell's statement was finally revealed to mean. Given that most industry analysts value oil companies on the basis of their proven reserves, it was a profoundly shocking message.

The financial wires and reporting services went into meltdown. In banks, boardrooms, newsrooms, bourses, investment offices and oil ministries across the world, there was incredulity and disbelief. Careful, ultra-conservative, Calvinist Royal Dutch/Shell caught by its own admission massaging the numbers? And by almost four *billion* barrels? It was like discovering your mother moonlighting as a pole dancer – not so much unlikely as unreal, surely?

Initial bemusement turned, in a market all too recently mauled by the

Enron scandal, into the kind of selling that Shell's statement made inevitable; the share price plunged 8 per cent, wiping £3 billion off the value of the company, in an hour of trading in London.

Explanations and answers were sought with increasing urgency from Shell's South Bank headquarters. And it was now that a palpable crisis turned into a major disaster. Instead of making available chairman Sir Philip Watts, head of exploration and production (E&P) Walter van de Vijver – the man most concerned with reserves – or finance director Judy Boynton, Shell elected to stage a teleconference and put up only hapless PR people to field questions.

There was anger and outrage among investment fund managers with the financial well-being of thousands of pensioners in their care, to say nothing of energy ministers whose national economies floated in no small measure on oil the company claimed was in reserve beneath their feet.

They – and thousands of investors great and small – wanted to know where, with Shell's blue-chip status going to hell in a handcart and the share price perched on the edge of a precipice, the company's chairman might be found. Sir Philip, a committed Christian and passionate plantsman, was, according to a report carried by BBC TV, at home in Berkshire giving his full and undivided attention to the Japanese garden he had created in the grounds of his Binfield mansion. He was not taking calls, nor were the rest of Shell's top brass.

The company said the downgrading in status from 'proven' to 'probable' of 20 per cent of its reserves followed a review which showed overestimates of 2.1 billion barrels in Nigeria and Oman, 1.2 billion in Australia and Kazakhstan and 0.6 billion in other fields around the world.

An industry commentator and former Shell manager commented to one of the authors, 'the fact that Watts was neither prepared to front the bad news nor appear in public to explain these numbers rang alarm bells.

'I was just one of many people who thought that Shell was only showing us the tip of a very large iceberg. We wanted to know what was going on under the water. We wanted authoritative explanations. The way the company handled – grossly mishandled, actually – the announcement encouraged suspicion and, by leaving major questions hanging in the air, invited speculation.'

And there was indeed much speculation about what had prompted both the review and such a drastic revision of the figures. For Shell, like all oil companies whose shares are publicly traded in the United States, has to register reserves in accordance with guidelines and criteria laid down by the Securities and Exchange Commission (SEC).

Reliable reserves numbers honestly filed are of critical importance to investors because the ability of an oil company to generate future earnings, and for how long, depends largely and obviously on the volume of resources available to it. To be sure, the classification and accounting of reserves tends to be arcane, but it is neither rocket science nor a novelty. The regulations have been in place for years and Shell, like all the other majors, has staff who specialise in reserves arithmetic and record-keeping. And anyway, you could be sure of Shell, couldn't you?

In truth, some analysts and industry insiders had been saying for years that Shell was vulnerable on reserves. A famously 'crude hungry' company with a history of transporting, refining and marketing oil found and pumped by other people, Shell – as always – had stood alone and aloof during the industry's most recent bout of consolidation, when giants such as ExxonMobil and ChevronTexaco grew even bigger and Total merged with Elf and Fina.

By merging and consolidating, the other majors – including BP with its acquisition of Amoco and Arco in America and Burmah Castrol in Britain – had been able to 'buy in' considerable reserves. Shell, despite bold claims to the contrary and the setting by Watts of some formidable and much-trumpeted overall growth targets, was in contrast reckoned to have seriously underperformed in the crucial reserves area.

The respected analysts Wood Mackenzie were quoted in *The Times* as saying that Shell's ratio of reserve replacement (RRR) for the period 1997–2002 was 111 per cent. That is to say, the company had added to reserves 1.1 barrels for every one it produced and sold. But slashing proven reserves by a fifth made nonsense of these figures. Take almost 4 billion barrels off the previously declared total and you reduce the RRR to 63 per cent. And that, Wood Mackenzie said, means Shell 'are finding less than two-thirds of what they produce'.

There were serious cost implications, too. Because fewer barrels met the criteria demanded of proven reserves, the average cost of finding and producing had soared from $4.27 of previous estimates to $7.90 per barrel.

Deutsche Bank had also been running the numbers in the wake of

Shell's startling statement. Reporting the Bank's similarly disturbing conclusions, *The Scotsman* said:

> [The company] has long been criticised for failing to replace the oil it pumps with sufficient new discoveries to maintain its production levels in years to come.
>
> In 1996, Shell said it achieved a replacement ratio of just over 200 per cent, but that slumped to 100 per cent by 1999, and fell below 50 per cent in 2001–02.
>
> In January, Shell revealed these replacement rates had been overstated. Its downgrades show that the ratio fell from only 160 per cent in 1996 to almost zero in 2001–02, according to estimates by Deutsche Bank.

The paper also noted that the SEC had begun an investigation into a Shell senior management bonus scheme that linked some executives' performance and pay to the company's RRR.

Despite its gentlemanly image and reputation for good corporate manners, Shell insiders say that over the years the company has developed a particularly vicious blame culture. This was put on public display on Shell's own Black Friday when the company's spinners sought to shovel wholesale responsibility for the reserves fiasco onto 'over-ambitious' managers of far-flung operating companies. Executives concerned with Australia's giant Gorgon gasfield – big enough on its own to meet world demand for four years – and Nigeria's prolific Delta oilfields were singled out for having booked 'immature reserves' in attempts to boost the performance and visibility of their own empires. According to the company's spokesmen, the errors were made in the mid-1990s and before Sir Mark Moody-Stuart, the former chairman, had in a series of moves aimed at strengthening the centre pruned the power of Shell's regional chieftains.

The business of booking reserves from Gorgon – a huge, £4.6 billion development in which Shell has a 29 per cent stake in partnership with operators ChevronTexaco (57 per cent) and ExxonMobil (14 per cent) – before a buyer had been found for the gas was the focus of much attention. For a move of this sort was plainly – glaringly – incautious and wholly out of step with the financial conservatism for which the company is renowned.

Maybe, said the spinners, but Shell had letters of intent from prospective buyers of Gorgon gas. The answer was dismissed with a derisive snort by analysts, who said letters of intent in this context were nothing more or less than an indication of interest in talking about possible terms and conditions.

There was similar scepticism over the 'losing' of a billion barrels from Nigeria, a matter reckoned by many observers to be even more remarkable. It was Shell's explorers who found oil in the country in the mid-1950s, since when the Nigerian operation has been of crucial importance to the company and the cause of great, continuing and damaging grief to the Ogoni people of the Niger Delta.

Given both the scale of the Gorgon project and the long history of Shell's familiarity with and involvement in Nigeria, observers said it was almost inconceivable that decisions – and errors – of such magnitude could have been made without some sort of headquarters oversight.

This was particularly true at a time when all oil companies were with increasing urgency trying to find and develop oil and gas in fields outside of the OPEC-dominated Middle East. For although Saudi Arabia, Iraq, Iran and the micro-states of the Persian Gulf hold the great bulk of the world's reserves, their governments are becoming increasingly reluctant to permit the rapid exploitation and equally swift depletion of their only tangible assets.

Moreover, sceptical observers pointed out that the company's insistence on blaming overseas operations for the reserves debacle obscured an important home truth and common thread. For the man in charge of Shell's business in Nigeria in the run-up to the 'losing' of the billion barrels was Philip Watts. And the company's head of exploration and production – and thus the man most intimately concerned with reserves on a worldwide basis – at the time when Gorgon, Oman and Kazakhstan became problematic was . . . Philip Watts. And the man most obviously missing from the Black Friday proceedings was, of course, Sir Philip Watts.

Not the smoothest of suits to take the top job at Shell, Watts – knighted like so many of his predecessors for services to British business and industry – is a tough-minded seismologist with a penchant for cost-cutting and zero tolerance of fools.

Born in Leicestershire in the English Midlands in 1945, he was educated at Wyggeston Grammar School and Leeds University, where, having achieved an honours degree in physics, he went on to gain an MSc in geophysics. Never universally popular within Shell, Watts has been described by colleagues as 'a hard man to know and a harder one to like' and 'something of a conundrum'. Already a convinced Christian, after graduating Watts spent a year teaching at the Methodist Boys High School in Freetown, Sierra Leone, before becoming a Shell 'lifer' in 1969.

Brusque, prickly and highly competitive, Watts was reckoned to be as tough as tungsten and his appointment was welcomed by big investors who had been less than impressed by a succession of 'Shell smoothies' in the top job. But Watts was said by friends not to be entirely lacking in humour and humanity: he had for many years been giving part of his – very considerable – earnings to his church.

There was, too, a just-detectable hint of humility in the fulsome apology Watts offered in response to the fierce criticism both he and finance director Judy Boynton were subjected to for failing to face journalists and analysts personally when the reserves fiasco was announced. Speaking on 5 February at the presentation of Shell's 2003 results, a chastened Sir Philip was reported by the *Daily Telegraph* as saying that the events of 9 January had been 'seared' into his memory. 'It was a mistake not to be there. I regret that and I am sorry that I was not there. That is an unqualified apology. I got it wrong. That is the reality that I have to face,' he said.

Sir Philip revealed that he had considered resigning. 'I came to my own personal decision that I should not do that. This thing happened on my watch. I am very determined to see it through this difficult patch.'

Reaction to Watts' comments was mixed. The *Independent*'s City commentator Jeremy Warner said that while the 'grovelling' apology had been gratefully received, Watts had still failed to explain why the 'calamitous' reserves overbooking had happened in the first place. His credibility 'is shot to pieces' and the reality is that 'a new dawn requires a new man'.

There was similarly little comfort for Sir Philip in the *Sunday Times*, where William Lewis wrote: 'The Shell of the future must look very different and it is far from clear Watts has the will or the desire to implement major changes.'

AFTERSHOCK (1)

Watts' attitude to change would soon be of academic interest only. For the first shot in what would quickly turn into a rolling barrage of scandal featuring fat cats at war, lies, cover-ups, deceptions and the destruction of records was fired in a report on 25 February in *Alexander's Gas & Oil Connections*.

'Shell failed to inform shareholders and US regulators directly that it was receiving incentive payments from the Nigerian government for booking oil and gas reserves,' *Alexander's* said.

Pointing out that half the 3.9 billion barrels cut by Shell from reserves concerned fields in Nigeria and Australia, the report added:

> Under Nigeria's reserves addition bonus scheme, Shell and other overseas oil majors received tax credits for each barrel of oil booked.
>
> Nigeria benefited from the arrangement by being able to demand a bigger output quota from OPEC and higher prices from international oil companies when it auctioned off acreage.
>
> The bonus scheme ran for nine years from 1991. It was scrapped in 2000 by the new Nigerian government of President Olusegun Obasanjo which is now trying to recover $600 million from Shell and other oil companies.

Alexander's added that 'An oil industry source said: "During the 1990s there was real competition for reserves booking . . . everyone got carried away in the gold rush. Some of those reserves were not only not provable, they were not realisable."'

A six-day peace followed the Nigerian revelations. Hostilities resumed on Wednesday, 3 March in a welter of headlines. 'Shell's top executive forced to step down' in the *New York Times* and 'Watts forced out in Shell reserves fiasco' in *The Scotsman* set the broadsheet tone and heralded events that would end in the total destruction of the oil giant's century-old reputation for probity.

The only surprise about Watts' departure was that he had managed to postpone it for so long. But the fact that he had, in the *Scotsman*'s felicitous phrase, 'been joined at the exit by Walter van de Vijver, chief executive of the exploration and production business' caused much comment.

The *New York Times* story, under a London dateline, said:

> The extraordinary shake-up at the British–Dutch giant, a company known for its conservative approach to business and its Byzantine corporate structure, comes amid a formal investigation by the SEC.
>
> But it was a review of the restatement [of the reserves] by Shell's audit committee and outside advisers that led to the ouster.

The paper quoted Deutsche Bank analyst J.J. Traynor as welcoming the moves but also asking whether they were the result of shareholder or credit-rating pressure 'or if there is real evidence of malpractice'.

Ivor Pether, a fund manager with Royal London Asset Management, an investor in Shell's British component company, said: 'We've seen some decisive action from the board, and one has to read that as positive. But I was surprised to see Walter [van de Vijver] go as well as Phil [Watts].'

Van de Vijver, at 6 ft 7 in. tall an unmistakable member of Shell's top management team, was not only widely believed to be next in line for the chairmanship but also the man who had pressed for a review of the way Shell estimated its reserves. And the precise nature of Walter van de Vijver's views on Shell, its reserves and the way matters were presented to shareholders and the wider world was about to become devastatingly public.

AFTERSHOCK (2)

Jeroen van der Veer, Shell's new chairman appointed in succession to the sacked Watts, had been in place for just five days when the *Wall Street Journal* rolled a grenade under his desk.

On 8 March, the paper described an internal Shell memorandum advising the company's top management, including van der Veer, in early 2002 of possible overstatements of reserves and warning of 'inconsistencies' in the way they had been booked.

Reporting from Washington the next morning, Stephen Labaton and Jeff Gerth told readers of the *New York Times* that Walter van de Vijver had written to van der Veer, Watts and finance director Judy Boynton advising them of huge shortfalls in proven oil and natural gas reserves two years before they were made public.

According to memoranda and notes of executive discussions, senior managers decided against disclosure to shareholders and investors in favour of devising and implementing what was described in a July 2002 document as an 'external storyline and investor relations script'. This would seek to minimise the significance of reserves as a measure of growth and of the company's strength.

The documents also made clear that reserves problems had been under discussion months earlier. In February 2002, a memorandum stated that one billion barrels of reserves were 'no longer fully aligned' with SEC criteria because of a change in interpretation by the regulator. Moreover, reserves totalling another 1.3 billion barrels were reckoned to be at risk because of doubts that they could be extracted before licensing agreements expired with three foreign governments.

By July 2002, detailed concerns were being expressed. A memorandum highlighted what was described as the 'major challenge' involved in maintaining proven reserves – particularly in 2002–03 – while simultaneously increasing production and containing costs. Technical and commercial constraints, the document said, together 'equated to a shortfall of 2–3 billion barrels' of proven reserves.

Chairman van der Veer declined to comment on any of the issues raised – including the possibility of misconduct on the part of either serving or dismissed staff – on the grounds that he was awaiting the outcome of the review that Shell had itself commissioned. He said the results of the inquiry would be made public.

If he was hoping for a period of quiet reflection uninterrupted by the noisy intrusions of the press, he was to be sorely disappointed. Within days came news that the Financial Services Authority (FSA), Britain's principal market regulator and watchdog, had joined its Dutch and American counterparts in probing Shell's dramatic restatement of reserves. At the same time, Aad Jacobs, chairman of the company's audit committee, confirmed that the London office of lawyers Davis, Polk & Wardwell were investigating the affair for Shell.

Next up, again within a matter of days, came the leaking and publication of more confidential documents demonstrating that Shell kept secret from the Nigerian government important details of its cuts in recorded oil and gas reserves in the West African state. The company was fearful that publication of 'too much' information would not only damage business relations with the government but also jeopardise negotiations with the authorities over $385 million in bonus payments.

The documents showed that Shell had much to keep secret about its Nigerian operations. At the end of 2002, the company recorded 2.524 billion barrels of proven reserves in the country. But following reviews and a tightening of guidelines, senior managers were told in a December 2003 report by Walter van de Vijver that 720 million barrels were 'non-compliant' and a further 814 million 'potentially non-compliant'. Only 990 million barrels were wholly and fully compliant with SEC rules.

Observers noted that in a further demonstration of duplicity, a promise made by Nigerian President Olusegun Obasanjo at a 2003 conference to publicly reveal his country's oil revenues was well received by Shell together with the rest of the industry. The conference was in part organised by Transparency International of Berlin and was attended by Chris Finlayson, chairman of Shell companies in Nigeria. Transparency is a current Shell buzzword and is much used in PR documents. Mr Finlayson warmly welcomed the president's pledge of openness. A month later, Shell's senior managers were recommending details of bonus negotiations and reserves problems in Nigeria be kept secret.

ANOTHER DAY, ANOTHER CRISIS

If Jeroen van der Veer had received a baptism of fire as chairman, Malcolm Brinded, Shell's new head of exploration and production, was also feeling the heat generated by bitter controversy and investor fury.

Brinded, a Cambridge engineering graduate and mountain-biking enthusiast with a reputation for solid professionalism in the oil business, had been in Walter van de Vijver's old job for only two weeks when he was obliged to declare: 'I was surprised and I was disappointed.'

He was telling the press and analysts, called to a conference on 18 March, of his reaction to the barely believable discovery that another cut in reserves – this one of less than half a billion barrels – would have to be made.

In stark contrast to the 9 January debacle when none of Shell's top brass was prepared to face the press and analysts when the first calamitous cuts were announced, the company this time mobilised all of its field marshals. Besides Brinded – who as head of exploration and production ('upstream' in industry jargon) ranks in the top three at Shell – there were at the conference no fewer than three chairmen: van der

Veer; Aad Jacobs, chairman of the Royal Dutch supervisory board; and Lord Oxburgh, non-executive chairman of Shell Transport and Trading.

There had, said Shell, been a 'reassessment' of the Ormen Lange field in Norway. This had shown that reserves had been overstated and would now be cut by 470 million barrels. Although the revision was, in the words of the *Daily Mail*'s Alex Brummer, 'a mere bagatelle' compared with the first monster write-down, 'it exposes fundamental systems weaknesses which are inexplicable in a company that has been around for so long'.

Lex in the *Financial Times* was no more forgiving. Shell's breach of the SEC guidelines was 'inexcusable' even if after decades with little change, the guidelines themselves were 'out of step' with the sophisticated seismic mapping technology now in use.

The market took a dim view of this second assault on reserves – made all the more embarrassing by the fact that the uncut numbers were quoted as definitive in the annual results published on 5 February – and shares dropped 3 per cent, knocking £1.5 billion off the company's capitalisation at one stage.

Beyond the initial wrath of shareholders large and small, and further heavy denting of Shell's already battered reputation for basic competence and honesty, the row over the latest cuts and revisions of reserves served to renew interest in precisely who knew what, and when.

According to a report in the *Sunday Telegraph* of 21 March, van der Veer had a few days earlier claimed not to have known about the false reserves bookings. 'Did I know about incorrect bookings [of reserves] in SEC returns? No,' he was quoted as saying. But in a confusing statement that smacked of casuistry, he also said: 'Until the end of 2003, we had documents in the company that were exposures to the SEC. The word "exposure" is not an incorrect booking.'

But some commentators, already impatient and rapidly becoming contemptuous of Shell, said that the blame game was for most investors a sideshow obscuring much more important issues about a major oil company that had plainly lost its way.

The former Shell manager previously quoted, speaking after the 18 March conference, said: 'When you cut to the chase, what the reserves fandango is really about is the company's shortage of oil and gas.

'Decoupling reserves bookings from the calculation of annual bonus payments to guys in the exploration and production division, as has just been announced, is obviously a good thing and should have been done sooner.

'Malcolm Brinded knows what he's about – as, indeed, did Walter van de Vijver – and says the company is in a turnaround situation over reserves. The fact, however, is that Shell is still only aiming at a 100 per cent RRR over the next five years, whereas the other majors will be on 130 per cent or more.

'We've heard a lot from Shell recently about transparency, sustainability, organic growth, capital discipline and "assurance" accounting procedures. We've heard a lot less about where the oil, gas and growth will come from.'

While the row sparked by the second reduction in reserves was rumbling on, Jeff Gerth and Stephen Labaton again added to Shell's misery with a detailed *New York Times* report on the company's problems in the Sultanate of Oman.

Through its Petroleum Development Oman (PDO) operating company, Shell has been active in the Sultanate for half a century and, after years of costly and frustrating exploration, finally discovered oil in the early 1960s and began exporting in 1967. Oman has never been in the big league of Gulf producers and its reserves, compared with those of neighbouring Saudi Arabia and the United Arab Emirates, are on the nugatory side of modest. But precisely because Shell, lacking a big, cheap-to-produce, onshore source of Middle Eastern oil, has so often in its history been a company hungry for crude, even small volumes such as Oman's have always been important. That, however, did not prevent Shell from overstating the Sultanate's reserves by a whopping 40 per cent at exactly the time the company was running into serious difficulties with its pioneering enhanced oil recovery (EOR) techniques.

Much of Oman's oil is locked up in geologically complex, small and scattered fields, and is thus by regional standards expensive to produce. To wring maximum performance and benefit from Yibal, the Sultanate's oldest, biggest and most prolific field, the reservoir was flooded with water under pressure to sweep the oil to 200 horizontal wells drilled during a major – and costly – hi-tech EOR project.

Production was indeed boosted initially. But in 1997, output from Yibal began declining at more than twice the regional rate. And worse, while Sir Philip Watts was talking up in a series of highly optimistic reports the benefits of horizontal drilling and other EOR techniques as 'major advances' capable of transforming the reserves picture, 90 per cent of what was coming out of the ground in Yibal turned out to be water.

Analysts say that a widespread failure of Shell's hi-tech oil recovery techniques to deliver the heavily hyped benefits could have serious and obvious implications for both costs and reserves. The Yibal problem manifested itself at a peculiarly difficult time for the company in Oman, because Shell was seeking to extend its licence substantially beyond its scheduled 2012 expiry. Shortly before Christmas 2004, the company announced that agreements had been signed with the government extending PDO's key Block 6 (covering central and southern Oman) concession for the next 40 years. That was the bland PR version of the result of protracted talks that insiders say featured 'a robust and vigorous exchange of views' about the relative merits of maximised short-run production and production rates calculated to maintain the long-term health of reservoirs.

THE HIGHER MUMBO-JUMBO

If hunger for crude has been a recurring theme in Shell's history, so too has the incubus of its weird corporate structure. Formed in a 1907 alliance between Royal Dutch (60 per cent) and Shell Trading & Transport (40 per cent), the company evolved into a complex monster with two HQs – in The Hague and London – two boards, a sort of 'Supreme Soviet' comprising managing directors drawn from the British and Dutch components, holding companies, satraps running the affairs of literally hundreds of operating companies scattered around the globe, joint venture companies involved in projects of awesome financial dimensions and, of course, the Group, in which all are lumped together but which does not exist as a legal entity.

Shell was for many years held up as a management role model and the company was a shining – and greatly envied – example of corporate cohesion. But that, say old hands, was when management was based on robust regional organisations, bullet-proof controls, clearly defined responsibilities and a well-understood system of checks and balances. However, in a process described by one senior manager as 'reform by meddling', the old system was dumped in favour of running Shell through major operating divisions. 'The trouble with that,' said the disgruntled insider, 'was that in good times, when oil prices soared and stayed high for a while, the rival empires waxed fat with staff on the "more troops, bigger army, more firepower" principle.

'When prices crashed – as in '86 and again much more recently – there was bloodletting on a fearsome scale. All sorts of people were put outside the door. No more jobs for life or any of that. Trouble was, we lost a lot of competency. Too many of the people who grabbed their pensions and headed for the exit were people who knew how to run a business and were good at it.

'We ended up with lots of essentially enthusiastic amateurs and chanters of management-theory balls – you know, the higher mumbo-jumbo of our time retailed by frightful pseuds with flip-charts – and a bunch of highly qualified specialists who, well-meaning and all that, couldn't have conducted a whelk stall as a profitable business between them.

'I know I probably sound like Victor Meldrew, but too bad. Control wasn't lost suddenly at Shell. It was surrendered bit by bit by people who should have known better and it took something like the reserves fiasco to demonstrate what had happened,' the senior (and now retired) man said.

But it took the *Wall Street Journal*, in an extraordinary 2 November 2004 article by Chip Cummins and Almar Latour, to reveal how deeply Shell became infatuated with New Age 'management-theory balls'.

The affair began in the 1990s when seemingly permanently depressed oil prices led to the wholesale abandonment of traditional management virtues such as general prudence, puritanical accounting, the rigorous application of 'fitness for purpose' tests to almost everything and a conservatism of outlook – and in particular to forecasting – born of almost a century of hard-won and sometimes bitter experience.

In their place, said the paper, came credulity-stretching exercises such as that in Holland in late 2000 when the chief of Shell's Dutch exploration unit asked his planners to deliver five-minute revue-style sketches of ideas for the discovery of oil and gas.

In one, a man ran across a stage stark naked in a bid to grab the attention of his boss. Another, said Cummins and Latour, featured a mock episode of the *Jerry Springer Show*, 'the incendiary daytime talk programme. A third, after a bit of fun and games, promised to extract large quantities of gas from seemingly declining Dutch fields.'

And there was more and worse. On other occasions, managers were told to shake their arms up and down in New Age 'energiser' sessions and, moreover, to stare into the eyes of colleagues while revealing their innermost thoughts.

Most telling of all, however, was the treatment of sceptics. All session participants were obliged to sit in a circle. An empty chair was placed in the centre facing the group. 'Anyone who wished to speak out against a plan by the group had to take the empty chair. The set-up was designed to nurture consensus by discouraging "blockers", people resistant to bold moves,' the *Wall Street Journal*, quoting participants, reported.

By now, Shell veterans – and engineers in particular – were trading grim jokes about collapsing standards. Young, newly recruited technical staff were referred to as 'Nintendo engineers' because, like most of their generation, they were more familiar with computer modelling than with real life.

AFTERSHOCK (3)

The full extent of the catastrophe brought on by Shell's cultural revolution had already burst with the brilliance of a star shell at midnight in the headlines of the newspapers of 20 April.

'Lies, cover-ups, fat cats and an oil giant in crisis,' said *The Independent*. 'Shell-shocked . . . how oil giant fiddled the figures' and 'Scandal that taints a famous name,' said the *Daily Mail*.

The papers were reporting the findings of the inquiry Shell had commissioned from Davis, Polk & Wardwell, as well as carrying news of the removal of Judy Boynton ('Boynton gets the boot as Shell lifts the lid on reserves row' was how a tabloid handled the story), the feisty, 49-year-old American finance director. Amazingly, the company also chose to announce another 500-million-barrel cut in reserves. This, the third revision, meant that Shell had slashed 4.87 billion barrels from its proven reserves total.

But on what was a truly terrible day for Shell, the worst news was the inquiry's revelation that a war had been raging between Sir Philip Watts and Walter van de Vijver from the moment the Dutchman became head of exploration and production and had access to the books.

Clearly appalled at what he found, van de Vijver began filing a string of extraordinary whistle-blowing emails to Watts – his E&P predecessor – that 'premature' and 'over-aggressive' bookings of reserves had created a false picture. The company, he said, was struggling to 'fool' the market.

Watts responded by pressuring van de Vijver, saying in May 2002:

> You will be bringing the issue to CMD [committee of managing directors] shortly. I do hope this review will include consideration of all ways and means of achieving more than 100 per cent in 2002 – to mix metaphors . . . considering the whole spectrum of possibilities and leaving no stone unturned.

By 22 October 2002, van de Vijver's exasperation was plain to see. 'I must admit,' he said in a message to Watts, 'that I become sick and tired arguing about the hard facts and also cannot perform miracles given where we are.'

On 28 February 2003, the Dutchman warned: 'We know we have been walking a fine line recently on external messages . . . promising that future reserves additions are expected in 2003 . . . whilst we know there is some real uncertainty about this.'

Van de Vijver's exasperation boiled over on 9 November 2003: 'I am becoming sick and tired about lying about the extent of our reserves issue and the downward revisions that need to be done because of far too aggressive and optimistic bookings.'

On 2 December 2003, the E&P chief received a memo from his own finance staff warning that the company must disclose to the market its need to reduce its reserves or be in breach of SEC rules. His response was damning: 'This is absolute dynamite, not at all what I expected and needs to be destroyed.'

With the publication of the inquiry's findings, Shell's humiliation was almost, but not quite, complete.

Standard & Poor's immediate reaction was to cut the company's credit rating and, on 24 August, the results of both the SEC and FSA inquiries were published. Both regulators found that Shell's reserves auditors had warned the company as early as January 2000 – four years before any public admission of error by any executive – that the figures might have been overstated. The reports confirmed that Shell had agreed to pay fines of $120 million (£67 million) to the SEC and £17 million to the FSA.

Jeroen van der Veer was quoted in the *Financial Times* as saying: 'The conclusion of the FSA's and SEC's investigations into Shell represents another significant step in putting the reserves issue behind us.'

The chairman's comment did not, of course, take into account the fact that Shell faces lawsuits – including a class action – in America, in which claims running into billions could be at stake. And there was no reference either to the anger caused by what Jeremy Warner in *The Independent* called 'the jaw-dropping' news that Sir Philip Watts would receive severance pay from Shell of 'more than £1 million for presiding over the worst crisis in the company's history'. Sir Philip, it was revealed, would in future have to get by on a pension of more than £584,000 a year. Walter van de Vijver, who in 2002 was paid more than £1.1 million, can also take comfort from a 'generous' pension.

To be sure, Shell's share price recovered quickly as, indeed, it was bound to given oil prices – driven by more threats of violence in Nigeria's Delta region, frequent disruptions of the supply from Iraq and much greater than forecast demand in both China and India – crossing the $50-a-barrel barrier for the first time in the last week of September 2004.

But the affairs of Cairn Energy, a tiny – by Shell standards – Edinburgh-based company, served in August 2004 to rub salt into the giant's still-sore wounds. For shares in the exploration firm trebled in value to £14 and its market capitalisation leapt to £2.1 billion on news of its fourth major oil find in Rajasthan since January. Cairn has so far discovered more than two billion barrels in just 10 per cent of a 6,000-square-kilometre block it bought from Shell two years ago for a knockdown price of £4 million.

Now, a major overhaul of Shell's reserves accounting procedures and, crucially, of its Byzantine structure, is being worked on by a top management team with radical changes expected to be swiftly implemented. But it will be many years – decades, perhaps – before Shell's wrecked reputation can be restored.

And it was that – 'our good name' – that mattered more than anything else to Marcus Samuel, the East End entrepreneur who turned a trade in seashells into one of the world's biggest, richest and most powerful oil companies.

PART ONE

From Shells to Hell

CHAPTER 1
FAMILY MATTERS

PICTURES IN AN ALBUM

First, some portraits from a Victorian family album. Photographed individually but clearly during the same studio session in 1860 are Marcus Samuel Snr and his wife Abigail.

Marcus, a merchant trader with a home in Finsbury Square and a shop, warehouse and office (M. Samuel & Co., established 1833) at 31 Houndsditch, is sitting, seemingly none too comfortably, looking well to the right of centre. Dressed in a severely cut, sombre coat and a stiffly starched shirt with a flyaway collar ready for take-off, he is plainly not enjoying the business of posing for the camera. Unsmiling and ramrod-backed, he looks tense, tetchy and very much as though he would rather be doing something – anything – else; the index finger of his left hand is fully extended and points, as though in mute accusation, at the floor.

Abigail Samuel looks, if anything, even less comfortable than her husband. The mother of seven children, five of whom survived beyond infancy, she is posed identically and also looks right of centre. She is wearing a voluminous dress with mutton-chop sleeves and skirts of marquee-like proportions.

And here, posed similarly in a studio session in 1871, are two of the Samuels' sons, Marcus Jnr, born in 1853, now 18 and recently returned from Paris and Brussels, and Samuel, who is 17. The boys sit with their legs crossed, Marcus left over right, Samuel right over left. Marcus has soft features, dark eyes and the look of a man who will soon be struggling with a weight problem. His curly black hair has been badgered by brushes into a sort of early Roy Orbison style; it looks artfully contrived and suggestive of more than a touch of youthful vanity. Samuel, by contrast, has been content for his hair to remain curly for the photographic record. Both boys have permitted themselves the faintest of shy smiles.

Exactly why Marcus Snr consented – his edgy distress and all too apparent distaste make it impossible to believe that he initiated the trip to the studio – to being photographed back in 1860 can now only be a matter of conjecture. But a celebration of some sort was surely merited because Marcus had in his business done very well for his family and himself. The descendant of Jewish immigrants from Holland and Bavaria who arrived in Britain in 1750, Marcus was described in the 1851 census as a 'shell merchant' and was listed elsewhere as proprietor of 'The Shell Shop' in Houndsditch. In truth, he had set himself up to trade primarily on the wharves of London's docks – many of which were within walking distance of both his earlier home, hard by the Tower of London in Upper East Smithfield, and his office – where he bought seashells and other eye-catching curios from returning seafarers.

Business was good because the shell-decorated trinket boxes Marcus sold were very much to the mid-Victorian taste for the exotically ornamental. This was the time when colourful creatures great and small were slaughtered by the arkload to be reincarnated, stuffed and mounted in glass cases in suburban sitting rooms. And there was also a huge, fashion-led demand for the ostrich feathers and peacock plumes that Marcus dealt in.

Marcus Snr was not a man to confine his business activities to keeping the likes of the Pooters supplied with fripperies and feathers. He travelled widely and, by Victorian standards, often. It was said you could number his journeys to the Far East and Japan by counting his children, each of whom was reckoned to have been conceived on his return from the Orient. By the 1860s, the dockside trader had become a merchant of substance whose imports now included metals – tin in particular – and who was also exporting machinery to Japan.

Perhaps the most valuable legacy the young Samuel brothers inherited from their father was a network of reliable agents in the Far East and the sound business relationships he had developed with the great trading houses of Hong Kong, Kobe, Bangkok, Manila, Singapore and Calcutta. The full worth of the network was soon to become apparent, because Marcus Samuel died in 1870 and left the business to his eldest son, Joseph, with the proviso that the two younger boys, if they wanted to, should also have a share in it. The reality was that Joseph, who had never demonstrated any flair for business, was happy to make way for Marcus – who had been working for his father since 1869 – and Samuel to take over completely by the early 1880s.

And it was a propitious time for two able young men of ideas and some capital – Marcus and Samuel had each been left £2,500 by their father – to be in business. For the world was changing, profoundly and at a dizzying pace.

Consider that the tenor of the times was being shaped by such diverse influences as Darwin's *Origin of Species* and Samuel Smiles' *Self Help*, both published in 1859. It is easy now to dismiss the importance of the Scottish doctor's hymn to the virtues and character-building qualities of work, but its impact on Victorian society was enormous and long lasting. Having been turned down by Routledge – who cited difficulties created by the Crimean War as the reason – Smiles found a publisher four years later for *Self Help* in the firm of John Murray.

In its first year, the book sold 20,000 copies. More than 55,000 had been bought at the end of five years, 150,000 by 1889 and an incredible 250,000 by 1905. The core of Smiles' gospel was that a poor man could remedy his poverty and improve his social standing by thrift, determination and hard work – an idea as radical as anything proposed by Darwin in a society stratified by a pernicious, iron-bound class system, still believed by many to have been divinely ordained and sanctioned. Moreover, Smiles insisted that 'the energetic use of simple means and ordinary qualities' rather than genius would be chiefly responsible for bringing about the transformation of society.

The message was universal in its appeal and *Self Help* found even bigger audiences overseas when it was translated into French and German, Italian and Danish, Arabic and Turkish, Japanese, 'several of the languages of India' and Dutch. And the vigorous employment of 'simple means and ordinary qualities' was a tenet a young Dutchman, destined to play a vastly important part in Marcus Samuel's life and business, was going to make even more famous.

The Samuel brothers – with Marcus the innovative, adventurous maker of deals and quick decisions always in the lead and Sam happy to play a supportive, but far from submissive, role – were hard working and self-reliant, and, from the outset of their careers, epitomised many of the qualities Smiles urged on his readers. But it was undeniably their abundant good fortune to have been embarking on business on their own account at the zenith of the nineteenth-century 'can do' culture and at a time when technological change was turning the lives of people and nations upside down.

Railways, the lustiest child of the great Victorian marriage of steam and steel, had by the last quarter of the century wholly revolutionised domestic travel and the economy. Now, express trains hauled by locomotives whose power was matched only by their functional elegance thundered through transformed landscapes towards cathedral-like termini such as Cubitt's King's Cross, Gilbert Scott's St Pancras and, of course, Brunel's magnificent Paddington.

And it was the incomparable Brunel, an arrogant, irascible engineering polymath of genius, who had brought a tidal wave of change to the merchant marine with his iron vessels of unexampled size, power and speed.

The *Great Western* was first. She crossed the Atlantic in 15 days and was by far the largest vessel then afloat. Next, with the *Great Britain*, Brunel made first use of screw rather than paddle-wheel propulsion. His third and final ship, the *Great Eastern*, was stuffed from stem to stern with what, at the time, were dismissed as technological novelties but which rapidly became standard features. The vessel was to remain the biggest in the world for more than half a century. Brunel did, however, have problems with all of his ships and many believed it was the number and magnitude of the difficulties with *Great Eastern* that broke his health and led to his premature death at the age of 53.

What mattered to merchant traders like the Samuels was that Brunel had effectively shrunk the world. And this, coupled with the opening in 1869 of the Suez Canal, which shortened the London–Far East route by 4,000 miles, and the development of wireless telegraphy, which reduced to seconds the time required to exchange information between Britain and India, China, Singapore, Japan and even Australia, had opened huge new opportunities for the entrepreneurially inclined.

And there was, too, the intriguing new business of oil.

THE COLONEL'S STRIKE

At the end of August 1859, 'Colonel' Edwin Drake, drilling at a depth of 69.5 feet in farmland at Oil Creek, two miles downstream of Titusville, Pennsylvania, found what he and his exhausted financial backers so earnestly sought.

Drake was an amiable sometime farmhand, railway conductor and odd-job man. His personal history was apt to be unpredictable since he was inclined to see a simple truth as a challenge to his powers of embellishment, and he had brought to his task no special qualification beyond an unshakeable belief in his ultimate success and the missionary zeal of the true believer.

He was not, nor had he ever been, a career soldier: his colonelcy had been conferred on him by his backers in the hope that letters thus addressed would give him some sort of credibility in the tiny Titusville community. Drake certainly needed all the help he could get in the matter of plausibility, since both he and his notion of drilling for oil in salt-boring country were widely believed to be crazy. Local sceptics, who turned Drake's endeavours into a spectator sport and gathered at the well site to sneer and jeer, missed no opportunity to point out that he had been labouring for more than a year with nothing to show for it.

And he would still have pitifully little to show for his pioneering work when years later and at a time when oil had already made some men rich beyond reckoning Drake would cut a peculiarly sad figure. For the man who had been born in Greenville in rural New York State in 1819 had after his Titusville triumph become an oil buyer and then a partner in a Wall Street firm trading in oil shares. But by 1866 he was penniless. Deeply ingrained improvidence and seriously unsound commercial judgement bordering on recklessness had quickly proved a disastrous combination.

In poor health, great pain and grinding poverty, Drake was obliged to plead his case in begging letters to his few remaining friends. Relief, of a sort, came in 1873 when the Pennsylvania state legislature granted him a small annuity. Towards the end of a life which at times had seemed like a mad parable, poor Drake tried to make sense of exactly what happened in Oil Creek on the fateful weekend of 27–8 August 1859.

After a fruitless year of hand digging and with financial support rapidly evaporating – he would soon be down to a single backer – Drake reckoned that success could only be achieved now by the sinking of a well. That, however, would entail hiring freelance salt borers, men skilled in the use of drilling equipment but also notoriously prone to lubricating their labours with prodigious quantities of hard liquor. Drake simply could not afford to put an itinerant bunch of drunks on the payroll.

In the end, financial stringency imposed a sensible solution on the problem: Drake himself erected the steam engine that would power the plant and hired a reliable crew, in the ample form of William 'Uncle Billy' Smith and two of his sons, to do the drilling. Uncle Billy was a blacksmith who made drill tools used by the salt borers and thus had some idea of basic procedures. Employed by Drake, he and his boys would be paid only by results based on the footage they successfully drilled.

Work began in the spring. It progressed with glacial slowness. The money, by contrast, disappeared at a gallop. With the hole stuck at the puny depth of 16 feet – Drake thought they might well have to drill 1,000 feet to find anything – and Uncle Billy in despair over the frequency with which the walls of the well collapsed, the venture seemed sure to end in ignominious failure.

Drake, however, refused to countenance defeat and, at this desperate juncture, conceived and introduced his innovative 'drive pipe'. A ten-ft length of cast iron in two jointed sections each of sixty inches, the pipe could be driven – battered even – into the hole and through rock with the bit or other tools working safely inside. It was a major advance but the enterprise remained plagued by problems.

With the drill bit progressing at the funereal pace of three feet a day, the steam engine which powered the rig burst into flames. As soon as the fire had been extinguished, the engine was rebuilt and restored to service. But it was surely in vain because the venture's sole remaining backer, James Townsend, a New Haven banker who had for some time been meeting all the bills from his own pocket, sent Drake a final payment and told him by letter to wind up operations in Oil Creek.

Townsend's instructions had yet to reach Drake when, late in the afternoon of Saturday, 27 August, the drill dropped into a crevice and then slid a few inches further. Work was stopped for the weekend but Uncle Billy went to view the well the next day, Sunday, and saw a black, viscous fluid floating on the surface of the water. He drew a sample and knew immediately what he had in his tin cup.

When Drake arrived next morning, he found his drill crew surrounded by tubs, buckets and barrels of oil. Drake went to the well and did what the sceptics and doubters – and they were legion – insisted was impossible: he began pumping oil out of the ground.

That same day, Drake received Townsend's letter and orders to close down. It was, of course, too late because the great oil rush – bigger, brasher, barmier than even the Klondike – had begun there and then in Oil Creek, Titusville. Drake had triggered an inescapable tsunami of change. Truly, nothing would ever be the same again.

Within 24 hours, derricks had been built and set to work as close as possible to Drake's site. And soon, all sorts of people would begin to make all sorts of claims to a kind of contiguous fame about the events at Oil Creek. Some, and not least those of the original backers who had tired of the venture and withdrawn their support, were bitter and vituperative. Drake, who had of course neglected to apply for a patent, was in pain and penury when in retrospect he claimed the invention of the driving pipe. He said he could also rightfully claim 'to have bored the first well that ever was bored for petroleum in America'. It had been no mean feat; it was no mean epitaph.

Drake died in Bethlehem, Pennsylvania, and was buried there in 1880. But that wasn't the end of the matter because 12 years later his body was exhumed and – perhaps in posthumous atonement for all the rig-site jeers and jibes – reburied in altogether grander surroundings in Woodland Cemetery. Today, young boys fish in a pair of placid Oil Creek ponds where 'the colonel' once drilled, visitors converge by the busload on the Drake Museum and oilmen pay homage to the pioneer driller at the Drake Memorial.

Belated gravitas and celebrity notwithstanding, Drake discovered neither oil nor drilling. Chinese salt miners were sinking boreholes to depths of 3,500 feet more than 2,000 years earlier using brass-tipped bamboo poles as bits. They also used oxen to power machinery that would look distinctly rig-like today. Nobody in the ancient world knew more about the business of drilling deep holes through rock than Chinese salt borers with their percussion techniques, and it was probably the Chinese who first noted that oil was frequently found in close proximity to rock salt.

But it was the Sumerians who, even earlier, in about 3,500 BC, were bonding bricks and waterproofing boats with bitumen, a crude oil component that sometimes seeped out of the ground. Bitumen was also used in ancient Egypt in the mummification of bodies; seepages of the stuff were reported in many other parts of the world.

The combustible nature of oil was harnessed by the Romans, who regarded the smoke produced by burning crude as an excellent pesticide, while in Cairo the pyrotechnic qualities of petroleum distillate were demonstrated all too dramatically when an estimated 60,000 gallons went up in a spectacular – if catastrophic – eleventh-century blaze. The Cairenes had been using oil to make bright-burning torches.

The terror weapon of choice on medieval battlefields was an incendiary, napalm-like mix of petroleum and lime called Greek Fire. Suitable for being hurled by the bucketful over fortress walls by trebuchets, packed into handy grenade-sized terracotta pots or fired from a pump in the manner of a primitive flame-thrower, it was in the age of the bow, battle-axe and broadsword the stuff of nightmares. Greek Fire, which when doused with water simply spread and burned more furiously, retained its gruesome potency until late into the fifteenth century, when it was superseded by the arrival in Europe of gunpowder from China.

Elsewhere, on the shores of the Caspian Sea, for instance, people who had access to oil were using it for entirely peaceful purposes, such as heating and, above all, lighting. Marco Polo noted in 1272 that almost all residents of Baku used oil lamps and that people travelled considerable distances to buy oil from merchants in the Caspian town.

In the Far East, Britain's first ambassador to Borneo reported at the end of the eighteenth century that almost 500 wells, all excavated by hand, were supplying the oil that millions of people were using for heating and lighting. But in Europe, other than in extremely localised areas and communities, it was a different and darker story. Well-boring technology certainly existed and in France, rotary drilling was used in 1784 to sink a well to a depth of more than 1,800 feet. Artesian wells of this type were nearly always sunk in pursuit of either salt or water. There were, certainly, a few known sources of oil – generally in seepage form – but they were either too small or too remote to be worth exploiting.

When oil was found – almost always accidentally – it was regarded in Europe as an irritating pollutant of no commercial value. Even in Britain, where the country's trailblazing industrial revolution had created an increasing demand for machine lubricants, apparently limitless reserves of coal and abundant water supplies would ensure that steam remained the driving force of the economy.

Domestically, the choice at nightfall was still between going to bed or lighting a candle or lamp made from an assortment of more or less combustible materials. Rags and reeds dipped in animal fat provided cheap though often foul-smelling and frequently smoky illumination. Tallow candles were a cleaner and far more efficient alternative but were prohibitively expensive. The danger of fire was, of course, huge and constant. Town gas, distilled from coal, became available in the nineteenth century, but, as its name implied, only in major towns and cities and then only to the well heeled.

In America, whale oil fuelled clean, bright-burning lamps but at a fierce cost. In 1819, a 1,200-barrel cargo of sperm whale oil landed by the *Essex* – the Nantucket vessel whose sinking by an enraged whale far out in the Pacific the following year inspired Herman Melville's *Moby-Dick* – sold on the quayside for $26,500, or slightly more than $22 per barrel. The price reflected the ever-increasing scarcity of the animals and the three-year voyages whale hunters often had to make to secure a commercially worthwhile cargo.

All over the world, daylight ended in darkness of a degree and on a scale unimaginable now. And it was this universal fact of life, and the almost simultaneous discovery by two scientists of a brilliant new illuminant, that drove the search for oil.

In 1846 in America, Abraham Gesner, a Canadian former doctor, geologist and twice shipwrecked horse exporter, succeeded in distilling 'illuminating oil' – kerosene – from bitumen. Shortly after, James Young, a Scottish industrial chemist working in England, distilled liquid paraffin from petroleum found in small quantity by a coal owner at the bottom of one of his mines.

Young quickly returned to Scotland, where he began his distillation process again, only this time from the far more plentiful oil shales. He applied for a British patent in 1852, two years before Gesner sought a US patent for the kerosene he described as 'a new liquid hydrocarbon . . . which may be used for illuminating and other purposes'.

By 1859, a kerosene works in New York City that Gesner helped establish was producing 5,000 gallons a day and a similar plant was in production in Boston. In total, it was reckoned that the market for 'coal oil' – as kerosene was often called – was worth $5 million and that some

30 companies were actively involved. In Britain, Young had pioneered a refinery business based – of course – on coal, and another in France, which refined paraffin from oil shales. The first shipment of kerosene from the United States arrived in Britain in 1861.

Thus, the demand for oil was well established, as was the ability to refine it, and, at the end of August 1859 in rural Pennsylvania, Edwin Drake had demonstrated how the demand could be supplied.

At a market price of $20 a barrel, Drake's oil strike triggered an immediate realisation among bright young entrepreneurs – to say nothing of the rapaciously greedy – that here was a business with the potential to yield unparalleled profits. Within days – hours, actually – of oil being pumped to the surface for the first time, outriders of the gaudy circus of wildcatters, moneylenders, property speculators, brothel-keepers, hot gospellers, lease lawyers and liquor merchants who would converge on the site of every early find poured into Oil Creek and Titusville.

By June 1860, there were more than 70 producing wells and at least as many dry holes in the area. None of the wells was anything more than mildly productive. For here in the Pennsylvania backwoods, archaic English common law, a degree of geological ignorance and much naked greed combined to ensure that the oil was exploited with breakneck abandon.

Regulated only by the 'rule of capture', owners of land above a common reservoir of oil were effectively encouraged not only to plunder the pool individually but also as quickly as possible. The result was often ruinous: with drills and derricks working literally within touching distance of each other, some areas of land became as punctured as pin cushions in the vicinity of producing wells whose performance was seriously affected. Rows between landowners, drillers and disobliged investors were thus frequent and explosive.

In fact, although some fortunes were made and a handful of backers would eventually enjoy returns of $15,000 for every $1 invested, many lost their shirts. Events at Oil Creek in the three years following the first find were to demonstrate two great truths about the new industry: that prices could crash – from the initial $20 a barrel through $10 in January 1861 to 50 cents in June and, incredibly, just 10 cents a barrel at the year's end – and that chaos was the natural condition of the oil business. It was an industry of crises and extremes, feast and famine, boom and bust.

What had knocked the bottom out of the market was the fact that production – boosted by the finding in April 1861 of the first gusher flowing at the rate of 3,000 barrels per day (bpd) – had rocketed from some 450,000 barrels annually to more than 3 million in less than three years. Available supply raced ahead of demand. The pendulum, however, was soon to swing the other way and demand and supply would again achieve some semblance of equilibrium, with prices rising first to $4 then, nine months later in September 1863, to $7.25 before reaching $13.75 at the end of the Civil War.

By then, oil had begun to saturate the social and commercial fabric of America. Hundreds of firms were involved in the industry in one way or another and, as a deeply conservative banker noted with bewilderment, many hundreds of thousands of sober, industrious men – all, doubtless, disciples of Samuel Smiles – now preferred to risk their savings in oil rather than entrust them to the lacklustre security of the savings bank.

There was nothing dull or monochrome about life and work in the oilfields. In the mid-1860s, parts of Pennsylvania were witnessing scenes of Hogarthian excess. The action had now transferred to the less than exquisitely named Pithole (aka Pithole Creek). The first well was drilled in January 1865. Six months later, four flowing wells were producing 2,000 bpd – a great deal of oil at the time and almost a third of the region's entire output.

The wells and the fortunes they were making for the few fuelled the dreams and avarice of the many. There was just one deeply rutted and often impassably muddy road to Pithole, 15 miles from Titusville. It was a battleground on which teamsters, their wagons heavily laden with barrels of oil, sought to drive their horses against an incoming tide of wildcatters, hopefuls and hucksters.

This was capitalism red in unregulated tooth and claw. Land values soared to insane heights. A farm classified as practically worthless in early 1865 was sold in July for $1.3 million only to be traded again in September for $700,000 more. All too plainly, Pithole was no place for the faint-hearted or those of a sensitive disposition. Expanding at a furious pace, the town at times appeared to be a bedlam adrift on a sea of speculation and truly terrible liquor. The awfulness of the spirits sold to desperate drinkers in Pithole achieved national notoriety.

And then the oil dried up. The law of financial gravity took hold of Pithole – by now a place of some significance with fifty hotels, two telegraph offices, a pair of banks, a post office and a newspaper – with a vengeance. The tract of land that in September 1865 had sold for $2 million was auctioned again 13 years later. It made $4.37.

But Pithole was to have a profound and enduring influence on the oil industry worldwide. The monopolistic and, at times, violent and abusive conduct of the teamsters on that narrow rutted road had spurred the search for a means of uncorking the transport bottleneck. Pipelines, at first mocked and dismissed as a joke, were quickly shown to be economically effective and capable of swift development. It proved surprisingly easy to run small pipelines from producing wells to larger gathering systems, which delivered oil to railheads. The point was not lost on the tall, lugubrious figure of John Davison Rockefeller, who was sometimes to be seen picking his way through the Pithole mud.

THE AMBIVALENT BAPTIST

Like Edwin Drake, Rockefeller was born and raised in rural New York State and, like Henri Deterding – a man he would hear described as 'the Napoleon of oil' – he developed an early passion for figures and an astonishing facility for mental arithmetic. Both Deterding and Rockefeller would leave school at 16 and both would become clerks and bookkeepers. A pious, solitary and ascetic young man, Rockefeller went to work for a grocery shipping firm in Cleveland, Ohio.

Possessed of blue eyes that precisely matched the frosty bleakness of his temperament, there was a pitiless, vulturine quality about a man whose first instinct on meeting a potential business rival was to seek out his jugular. Throughout his long life – he was 98 when he died in 1937 – Rockefeller believed that he acted 'fairly'. Yet in amassing a colossal personal fortune as founder and supreme boss of Standard Oil, he laid about competitors with a ruthlessness that Vlad the Impaler would have recognised and saluted. In truth, tolerance and magnanimity were New Testament abstractions of limited appeal to a man who conducted his business affairs according to Old Testament rules; Rockefeller, tagged 'a bloodless Baptist bookkeeper' by an unimpressed Pithole wildcatter with a flair for alliteration, routinely put rivals to the sword. Competitors would be annihilated, crushed by the juggernaut of his implacable will.

An austere man of essentially puritan cast of mind who abhorred ostentation and set great store by personal rectitude, Rockefeller was also ambivalent and proved perfectly capable of stepping into the gutter to buy influence and favours. Seen outside his New York offices in Broadway at the zenith of his career in a black suit, bowler hat and velvet-collared overcoat, he looked for all the world like an undertaker. But this was also the man about whom the great muck-raking journalist and campaigner Henry Demarest Lloyd said: 'Rockefeller's Standard Oil could do anything with the Pennsylvania state legislature except refine it.' The legislature was the august institution that wags described as 'the finest body of men money could buy', and every man and his brother knew that the state government had been bought and paid for by three great tycoons – Rockefeller (oil), Carnegie (steel) and Frick (coal) – and the Pennsylvania Railroad Company.

At home in New York, Rockefeller was described as living like a frugal Scandinavian monarch. He slept with a bible always within reach at the bedside; hated, reviled and feared as he was by large numbers of his fellow men, he often sought solace in the scriptures. Frequently depicted as a cold-blooded being whose natural habitat was an emotional wasteland, Rockefeller took comfort from the notion that the staggering growth of Standard's profits was evidence of righteousness rewarded. If he was not directly engaged in the Lord's work, he believed that at the very least his enterprise enjoyed divine endorsement. To this singular tycoon who was also a Cleveland Sunday school superintendent, there was a mysterious, metaphysical quality about oil that transcended mundane issues of utility and profitability. Refined into kerosene lamp fuel, oil was a beacon capable of guiding society to a brilliant new era of enlightenment, stability, order and harmony.

These were precisely the qualities that Rockefeller found wholly absent from Pithole. Instead, chaos, instability, disharmony and disorder seemed to prevail.

He had pitched up in the oilfield because a group of Cleveland money men had asked him, a year after Drake's strike, to go and look at and report on the long-term possibilities of the news-making gushers. Rockefeller was viewed as an expert in the field and backed a refinery started by his friend Samuel Adams in 1862. He was not well received by the wildcatters, who soon discovered that beneath Rockefeller's icy exterior lay a comfortless core of permafrost.

Having completed his survey, Rockefeller returned to Cleveland, where, in another startling demonstration of ambivalence and of the profoundly different standards of conduct he believed to be acceptable in commercial life, he lied to his commissioners. The Cleveland investors were told that there was no future in oil.

That Rockefeller believed exactly the contrary is evidenced by his swift return to Pennsylvania, where, with his business partner Maurice Clark and very little capital, he opened a small refinery. For the first three years it was hard pounding. Often red-eyed from lack of sleep, he learned the business from first to last. He and Clark then took in three more partners and, over the next three years, made a thumping profit of $100,000. The refinery was put up for auction and, in what he would later describe as the most important business move he ever made, Rockefeller bought out all the partners for $72,000. At the age of 26, he was already modestly rich, greatly feared in the petroleum business and about to achieve legendary status as he created the mould that would shape all other oil companies and an entire industry.

Now out on his own, Rockefeller began by declaring war on everybody else in the business, beginning with fellow refiners. The measure of his success is the fact that by 1879, Standard Oil – the company he had founded nine years earlier – controlled 90 per cent of America's refining capacity and most of the pipelines (Rockefeller called them Standard's 'iron arteries'), and commanded the oil transport system.

What set Rockefeller apart from the horde of hopefuls who daily descended on Pithole was his ability to see the big strategic picture – where the industry was now and where it was heading – coupled with the iron grip of the tidy-minded bookkeeper on the mass of day-to-day detail.

It was this rare combination of talents that enabled Rockefeller to quickly identify a great central truth of the oil industry: size matters. If big was good, biggest was best in what was obviously an uncertain business. How could it be anything else given that drilling for the raw material was conducted on little more than a hope-and-hunch basis? Supply, clearly, was in these circumstances always going to be erratic and, to a large extent, uncontrollable.

What made matters worse and even more prone to disorder was the fact that refining/distilling had proved relatively cheap and easy to get into. Small, often inefficient and sometimes downright dangerous refineries were springing up everywhere. Former moonshiners – Henry

Flagler, the man who was at work in Ohio at 14 and would become Rockefeller's closest friend and, in 1867, partner, had already made and lost a fortune from whiskey distilling and salt mining when he helped form Standard – were nearly as numerous as more formally qualified chemists.

And because they were too many and, like the oil distributors and pipeline proprietors, too small, they spent far too much of their time hacking at each other in futile – though greatly damaging – competition. Rockefeller looked down his long hooked nose and viewed their cut-throat antics with repugnance. Contemporaries said he reserved special scorn for men too stupid to see that in seeking to open their rivals' veins they were bleeding themselves and the industry to death.

To Rockefeller, the solution was simple: combine under Standard's direction and ownership. The first great exponent of vertical integration argued that nothing else would produce the long-term stability and order essential for profitable growth. Economies of scale would drastically reduce all costs, but particularly in transport, where combination would permit refiners and distributors to wring vastly better rates out of the railroads. Oil bought cheaply during production gluts could be refined, transported, stored and finally marketed in a controlled, orderly and profitable way rather than released in a flood which drove prices through the floor.

This, said Rockefeller, would be better for everybody, including the consumer. It was to prove sensationally good for Standard in general and Rockefeller in particular because in a further, secret, refinement of the plan for dealing with railroad transport, he had organised an exclusive rebate scheme for Standard. This meant that all his company's oil would be carried at greatly reduced rates and some, indeed, would be transported free. The stratagem demonstrated perfectly how much better it was to be big in the oil business because as volumes increased, so did the scale of the rebates and the benefits to Standard. Even today, some important details of the notorious scheme remain secret. All that can be said for sure is that the rebates helped make Rockefeller even wealthier.

And the process of extraordinary enrichment – through the acquisition, with characteristic ruthlessness, of oil-producing companies and land, greatly increased sales of kerosene and lubricating oil, and constant, fine-focused attention to the control of costs – proved as unrelenting as the drive for improved efficiency.

But if Rockefeller was winning the balance-sheet battle, he was losing the war of public opinion. Already the subject of almost daily attacks in the press – Ida Tarbell most famously followed up what Henry Demarest Lloyd had begun in the *Chicago Tribune* – and by politicians in the wake of exposure by a New York State inquiry into the rebate scheme, he would, following a Congressional probe into the infamous Standard Oil Trust, become the most execrated businessman on earth.

Even though he seemed to live deep within an ironclad carapace that was utterly impervious to criticism, Rockefeller *did* care about his image – enough anyway to hire Ivy Lee, who, as the world's first PR practitioner, sought to radically change the popular perception of Rockefeller as a sort of Scrooge–Genghis Khan composite. Part of Lee's grand plan involved the commissioning of an official biography, and several high-profile specialist writers were considered. Among them was Winston Churchill, who said he would write the book for a fee of $50,000, an offer Rockefeller felt able to decline. The man who by the time of his death would have given – often without strings of any kind – $550 million to good causes thought better value might be obtained elsewhere. In the end, Allan Nevins undertook a two-volume study and made clear in the preface that he had received no special payment from its subject.

Through it all, and even at the height of the season of fear and loathing, Rockefeller's fortune continued to grow exponentially. What had begun as a small refinery business in Pennsylvania had metamorphosed into a predatory monster. Having launched an especially vicious price-cutting war – Standard's favoured tactic for breaking a competitor into bite-sized chunks – aimed at gaining a substantial measure of control over the Russian oil market, Rockefeller's interest settled on a British company caught up in the fray.

In 1901, he would eventually sanction the offer of $13 million to Marcus Samuel personally, a seat on the board of Standard and a total of $40 million for his business.

EMPIRE OF THE SON

It was in 1886 that Marcus and Samuel Samuel took their first tentative steps in the oil business. They bought kerosene in small quantities from

Rockefeller's Standard Oil and also from the great Far Eastern trading house of Jardine Matheson for resale in Japan.

By now, the brothers had two businesses: M. Samuel & Co., run by Marcus from his father's old Houndsditch premises, and Samuel Samuel & Co. in Yokohama. Sam, at the age of 23, had gone to Japan to establish the business. He would stay there for ten years.

The businesses were flourishing and were driven, in the main, by Marcus's energy – he had made two big tours of the Far East between 1873 and 1877 to maintain good relations with the network of agents and contacts established by his father – his flair for rapidly converting opportunities into deals and his ability to think on his feet and make quick decisions.

At the heart of the brothers' success was their expansion of the profitable two-way trade with Japan begun by their father. He had sent the first mechanical loom to the country and his sons would play an important, and acknowledged, role in the development of Japan by increasing the volume and variety of industrial plant, machine tools and textiles brought in from Britain under the direction of Sam in Yokohama.

From Japan, the brothers took rice, coal, silk, lacquer goods, copper and china. Under the sharp eye of Marcus in London, they also traded worldwide in foodstuffs such as sugar, flour and tapioca. And they, of course, continued to deal in the seashells – at one stage employing 40 women shell cleaners who carried out their work in premises rented in Wapping – on which everything had been founded.

It was good, profitable business and the brothers were already financially secure – indeed, modestly rich – by the time they were 30. But this was an important period in the life of Marcus Samuel for more than business reasons. He had, at the age of 28, married Fanny Elizabeth Benjamin, by whom he would have four children: two boys and two girls.

Always softly spoken and, in contrast to the bustling brio and ebullience of Sam, a complex, thoughtful man capable of subtle responses, Marcus seemed now sometimes slightly distant, a little withdrawn. The truth was that he had begun to nurture some serious social ambitions.

At a time when the kerosene strand of the business had grown rapidly to the point at which the brothers were considering shipping it, in a vessel of their own, as case oil (because of handling and safety problems, kerosene was carried at sea in cases each holding two five-gallon tins), Marcus was looking for more than money as a denominator of success. He wanted respect, kudos, status and position, and not only for himself but for his family, too. Above all, he yearned for acceptance.

Marcus knew only too well that the ordinariness of his background would make advancement difficult in a society as class-conscious as Britain's. In the United Kingdom in 1873, when the population totalled 32 million, an elite of just 7,000 still owned more than 80 per cent of the land. This tiny but vastly wealthy minority also dominated politics and government, and dictated tastes in the arts and even fashion. Two great Reform Bills might have established the supremacy of parliamentary democracy over the power of the monarchy, but the aristocrats and gentry remained the ruling class. And for Marcus there was another barrier on the road to personal progress: an East End boyhood in the shadow of the Tower was as nothing when set beside the vastly greater handicap of his Jewishness.

Anti-Semitism has a long and bloody history in Britain. There were serious disturbances in 1144 when, in the first of many accusations of ritual murder, Jews were said to have killed a small boy – remembered later as St William – in Norwich. Severe restrictions, including confinement to Jewries and later the mandatory wearing of large identity badges, were imposed but did nothing to either contain or defuse dangerously high levels of tension.

Following explosions of horrific violence, including, in 1278, a savage attack in which hundreds were hanged, all Jews were expelled from Britain. From then until the mid-1650s, there were no communities of any size and only a few individuals, most of whom professed conversion, slipped in.

Cromwell was more sympathetic and employed Jews in espionage and diplomacy, and, though there was no formal reversal of policy, their position gradually improved. It was the financial and commercial revolutions of the eighteenth century which enabled the Jews to increase both their status and numbers, and, as we have seen, the forebears of Marcus and Samuel arrived in Britain from Holland and Bavaria in 1750. Nevertheless, the old enmities, suspicions and hatreds proved deeply resistant to reform. When even a minor measure aimed at easing the naturalisation process was brought in during 1753, it provoked such uproar that it was immediately repealed.

By 1829, following the great parliamentary battles over Catholic emancipation, the Jews were the only faith group still suffering severe discrimination and disabilities. There were attempts at ameliorating their condition but they crashed on the rock-like resistance of the House of Lords. Yet Jews were beginning to make palpable progress in society. David Salomons became London's first Jewish Lord Mayor in 1855, having been Sheriff 20 years earlier, and Francis Goldsmid was the first practising Jew to be made a baronet in 1841.

But the continuing cruel anomalies and absurdities of the Jews' position were laid bare first by Lionel Rothschild in 1847 and then in 1851 by David Salomons. Both were elected Members of Parliament only to be denied their seats because, as non-Christians, they were unable to take the oath. The law was eventually reformed in 1858 and in 1868 Benjamin Disraeli became Britain's first Jewish Prime Minister. Disraeli, ethnically a Jew but a practising Anglican, had been a Conservative MP since 1837 and, because of his professed Christianity, had been neither discriminated against nor affected by the empowering reform of 1858. And it was Disraeli who in his 1844 novel *Coningsby* presented a sympathetic portrait of Lionel Rothschild in the character of Sidonia. Rothschild was to make another literary appearance, although of an altogether less flattering kind, when in 1874 he was thought by many critics to have served, at least in part, as the model for Melmotte, the great swindler in Anthony Trollope's *The Way We Live Now*.

By the time Marcus was seeking preferment and the most glittering of civic prizes – the Lord Mayoralty of London – public hostility towards Jews was rising to a new height. For between 1881 and the outbreak of war in 1914, a 150,000-strong wave of immigration would double the size of Britain's Jewish population. There was widespread alarm and dismay as Jews fleeing persecution in Central and Eastern Europe converged on British cities – London in particular – creating new ghettos, some of which quickly became no-go areas for gentiles.

The resulting tension was in no way assuaged by the virulent anti-Semitism of sections of the press or incendiary speeches by MPs such as the Conservative Major Evans Gordon, who complained that his Stepney constituents were in constant danger of being pushed from their homes by 'the off-scum of Europe'. But beyond the blind prejudice and orchestrated rage of the mob, there were legitimate causes for concern. Among the immigrants were criminal elements who quickly established prostitution, extortion and loan-sharking rackets which the police seemed unwilling to investigate and unable to stop. Moreover, in east London in particular, the new arrivals were in 1887 competing for jobs, undercutting wages and exerting a virtual stranglehold on some trades at a time of locally high unemployment.

VISIONS OF HELL

Marcus Samuel was a man with all this and a great deal more on his mind when, in 1890, he arrived in the capital of the Caspian oil region.

Baku was both a boomtown and a fiery vision of hell. Oil had been extracted and traded on the Absheron Peninsula for more than 2,500 years and Alexander the Great, during his 331 BC campaign against the Persians, used clay 'fire vessels' filled with Baku oil to light his tent. The district, with its 'eternal pillars of fire' – gas flaring from fissures in the oil-bearing limestone rock – was at the ancient heart of the Zoroastrian fire-worshipping religion. There were three Temples of Fire in the region at Surakhany, Pirallahy and Shubanu.

The area had a history of advanced well-digging, too. In the Balakan district, a shaft more than 115 feet deep was excavated in 1594 by Mamed Nur-oglu, 'a skilled man', according to the inscription carved into a large stone found at the bottom. Wells such as this were visited and recorded in the 1770s by a Russian scientist named Gemlin and were further researched and described in 1827 by Mine Engineer Voscoboynikov.

And it was in the Bibi-Eybat oilfield in Baku that the world's first oil well was drilled. According to Natig Aliyev, president of the State Oil Company of Azerbaijan, the well was drilled at the instigation and under the direction of a Russian engineer named Semenov in 1847.

In fact, an oil industry of sorts was up and running when the region was annexed to the mighty Russian empire in the early nineteenth century. After more than 50 years of maladroit and often deeply corrupt control as a state monopoly, oil exploration and production were effectively deregulated and opened to private enterprise in 1872. In a weird mirror image of events at Pithole, there was along the shores of the Caspian a stampede of entrepreneurs, an explosion of drilling activity and much furious construction of plant; by the end of the following year, no fewer than 20 small refineries were at work producing kerosene.

But only a missionary oilman with a deficient sense of smell could have found beauty in Baku as development raced ahead. The area seemed now to be thickly forested with derricks jostling to find a secure footing in what was often an ocean of treacly mud pockmarked by pools and puddles of crude sludge. And to be downwind of Baku was an olfactory experience like no other. Visitors noted that where once pristine 'pillars of fire' burned beneath a clear sky, the flames danced now under a reeking canopy of greasy, soot-laden smoke from the refineries. Workers and speculators flooded in; raw sewage flooded out from their haphazard camps and settlements to add to the noisome atmosphere. A classical fire-and-brimstone vision of hell had been swapped for a newer and much nastier version.

Production from the new wells – by the turn of the century there would be 3,000 of them with 2,000 producing oil in industrial quantities – rocketed from an annual rate of 600,000 barrels in 1874 to a thunderous 10.5 million ten years later. The total was almost a third of contemporary US production, a fact that did not escape the attention of John D. Rockefeller, who regarded the vast Russian empire as an immense market opportunity.

American kerosene had been on sale in St Petersburg since 1862 and US trade officials – some of the less fastidious of whom were augmenting their income with a little light forecasting for Standard on Rockefeller's payroll – were predicting big increases in exports in the near future. But Rockefeller's hopes of turning prognostications into burgeoning profits were to go up in smutty Russian smoke.

The main men in the booming Russian oil business were the Nobels – Robert and Ludwig, brothers of Alfred of dynamite and later Nobel Prize fame – and the Rothschilds, less a banking dynasty than a European financial superpower based in Paris.

The Nobels, through their Swedish emigrant father Emmanuel, had been in Russian business and industry with mixed fortunes for almost half a century when an impulse buy propelled them into oil. Robert, the eldest and least successful of the boys, had been despatched from St Petersburg by Ludwig to buy timber from which rifle stocks could be manufactured for a major arms contract their company had won. Instead of wood, Robert, travelling through the Caucasus, bought an oil refinery he had been offered in Baku. It was a poor, rundown and inefficient specimen but Robert, a chemist by training, reckoned that given some investment capital, it had commercial promise.

Robert immediately set about justifying his purchase – made, of course, with money provided by Ludwig – modernising and expanding the plant into an efficient model of its kind. When Ludwig visited the refinery, he saw at once the enormous potential of the Russian oil business. For Ludwig was a visionary with a phenomenal, Rockefeller-like gift for detailed analysis and swift mastery of detail while still retaining a clear view of the big picture.

Soon, Ludwig – who also had excellent contacts with the Imperial government – had expanded the interests of the Nobel Brothers Oil Extracting Partnership into every important sector of the industry. An energetic innovator and an organiser of genius, Ludwig rapidly began bringing order out of chaos in Baku.

For years, oil pumped from wells had been stored in a variety of pits, trenches and troughs ranging in size from modest ponds to great lakes. Huge amounts of oil were, of course, lost to evaporation and, worse, percolation of the puddled earth banks and bottoms of the retaining ponds. Groundwater in the vicinity of the pits must have been amongst the most contaminated in the world until Ludwig introduced his revolutionary iron reservoirs. Soon, there were huge tank farms all over the oil region.

A much bigger and more intractable problem was posed by Baku's geography. Already refining 50 per cent of all Russian kerosene, the Nobels were as dominant in their domestic market as Rockefeller was in his. But for Ludwig, transporting oil was a nightmarishly difficult and expensive business.

Baku was surrounded on all sides by barriers. To the west lay the impassable Caucasus mountains more than 16,500 feet high. They completely blocked the route to the Black Sea, which, 500 miles away, was the key to accessing the Mediterranean. To the south, east and north was the landlocked Caspian. Getting oil to the Baltic, the nearest open sea, meant that every cargo intended for export had to travel 2,000 miles before it exited Russia.

There was also the fearsome cost of barrelage. Because there was little or no suitable timber for local manufacturing, the wooden barrels used for exporting kerosene had to be imported ready-made, adding even more to an already prohibitively high transport bill.

Ludwig's first essay in cost reduction was the construction of a pipeline between the Nobels' wells and their refineries. The distance was short but the savings – in horse and cart journeys – were disproportionately large: the venture paid for itself within a year.

Next came a vastly more ambitious scheme – the design (by Ludwig) and construction at the Swedish Motall yard of a small steamship named the *Zoroaster*, which, after initial modification, became the forerunner of all modern oil-carrying vessels in which the hull is formed by the tank.

Operating in the relative calm of an inland sea, the *Zoroaster* was such a success that the Nobels built a fleet of similar vessels with names such as *Darwin*, *Moses* and *Spinoza*. The tankers boosted Caspian trade to such an extent that by 1890 Baku had become one of the world's busiest ports.

With the Nobels firmly in control of the route to the Baltic, the independent oil producers of the region were looking for help. It arrived in the form of a railway, funded by the Rothschilds of Paris, which went from Baku through the Caucasus to Batum on the Black Sea. Though formidable gradients meant that trains had frequently to be broken into short 10–12 wagon sections for haulage up the severest climbs, the impact was nevertheless immediate and galvanic: Russian case oil could now compete directly with Standard's in Europe.

Confronted by a serious business challenge, Rockefeller – as always – responded by putting his rivals on the rack of a price war. And, as always, price cuts in target markets were subsidised by customers who had the misfortune to live in an area where Rockefeller's monopoly gave them the choice of paying sharply more for their oil or going without.

But Rockefeller could see that to compete effectively over the longer term, he would have to reduce the cost of shipping his oil across the Atlantic. Thus, in 1885, Standard's German subsidiary commissioned a new ship, the 2,975-ton SS *Glückauf* (Good luck). Built in Britain, she was the world's first genuine ocean-going tanker, capable of carrying bulk kerosene from the US to Europe about 25 per cent more cheaply than a comparable cargo of case oil. Launched in 1886, the *Glückauf* was the first of 80 similar vessels which, within 10 years, would be engaged in the Atlantic oil trade.

Though offering significant savings to operators, the ships were detested by crews and proved only marginally more popular with insurers. In heavy seas, the vessels were unstable and remaining afloat was often more a matter of luck than maritime judgement. Moreover, explosions were depressingly familiar; the tankers were given a wide berth by other vessels in ports on both sides of the Atlantic. But the *Glückauf* – or rather the concept of a large-capacity bulk oil carrier – impressed and greatly interested Marcus Samuel.

Just when it seemed that the Russian kerosene business was as competitive as it could be, the Rothschilds rolled a grenade under all the other players by announcing in 1886 that they would be going into the trade with their Caspian and Black Sea Petroleum Company. They would, they said, use the railway they had financed to undercut both the Nobels' and the Russian independents' oil. The Rothschilds' company soon became known by the acronym Bnito, its initial letters in the Cyrillic script. And Bnito was about to become of critical importance to Marcus.

The Nobels, now under the competitive cosh from all directions, responded with some truly audacious engineering aimed at slashing transport costs to the minimum. They constructed a pipeline from Baku to the Black Sea port of Batum, a distance of 552 miles. With 16 pumping stations en route, it was in 1887 the biggest project of its kind ever committed to construction. But more impressive than its overall scale was the fact that 42 miles of the pipeline were laid in a tunnel blasted through the heart of the mountains. More than 400 tons of explosives were used: having a brother who not only invented dynamite but manufactured it too was clearly no disadvantage. Unhappily, Ludwig did not live to see the work completed. He died of a massive heart attack at the age of 57 while on holiday in France in 1888.

The net result of all the developments was a significant gain in world market share for Russian oil. In 1888, the US supplied 78 per cent of all exported kerosene. Russia's share was just 22 per cent. Three years later, Russian kerosene accounted for 29 per cent of the total and America's market share had fallen by 7 per cent. In terms of both volume and monetary value, the changes were of huge and lasting significance.

But they were trivial when set beside the revolution Marcus Samuel had in mind as he travelled towards Japan. He had been to Baku to see the Russian oil industry for himself because the Rothschilds, through Bnito, were planning to export Russian kerosene to the Far East in head-to-head competition with Standard Oil. And the Paris-based bankers reckoned that in Marcus Samuel – first recommended to them by Jardine Matheson – they had exactly the right man to execute their scheme.

The Rothschilds' approach to Marcus had been made by Fred Lane, a big, heavily built workaholic whose appearance belied an acute and agile business brain. Lane was senior partner in Lane & Macandrew, the shipping brokerage Marcus and Samuel Samuel used when they chartered vessels for their merchant trading businesses. But Lane, a hugely respected City figure and an important player in the Far East trade, was also known as a discreet intermediary and matchmaker of impeccable judgement. Capable of arranging the happiest and most felicitous of business marriages, Lane often invested his own money in the arrangements he brokered.

As Marcus neared Japan and meetings with some of his key contacts, he knew he was on the brink of a momentous, life-changing decision. Everything – hopes and fears, possibilities and probabilities – had to be weighed in the balance.

Now 38, he was a man of substance – the brothers had netted £1 million by buying rice cheaply in huge quantities during a glut and selling it when an Asian drought sent prices rocketing – and within touching distance of achieving the recognition and acceptance he so earnestly sought. He had already begun campaigning – discreetly, of course – to become Alderman of the Portsoken ward of the City of London. The post, although largely honorific, mattered because the Lord Mayor of London was always selected from among those who had previously held aldermanic rank. And even if the business of being Lord Mayor had a risible side – pantomime wigs, tricorn hats, silk breeches, a chain of office that would not have looked out of place round the neck of a dray horse and a bejewelled mace of imperial proportions – it was nevertheless the highest civic post a man might aspire to in Britain. Also, only two Jews had ever made it all the way to the Mansion House and the summit of City government.

There was, too, the business of moving to a larger, grander house in Portland Place. Although within a few miles of his birthplace on the map, it was light years distant in terms of neighbourhood. Later, there would be the acquisition of land on a grand scale with a vast country home, called The Mote, sitting in its own 500-acre estate at Bearsted in Kent.

And there was the question of the boys' education. There would be no learning on the hoof in Europe for them. Marcus had both of them down for Eton, the most illustrious of English public schools, where they would have as classmates the offspring of foreign royalty and more future British cabinet ministers than you could shake a stick at. Then, as now, you needed deep pockets to send your boys to Eton.

With Samuel back in London permanently – and destined to become a Conservative MP – the business had forged ahead. The brothers, despite their wholly different personalities, had formed a powerful and effective combination. Their relationship in the office – Marcus, who had a horror of overheads, had retained the Spartan accommodation favoured by his father in Houndsditch – was volatile and punctuated by rows and storms, but neither man bore grudges and equilibrium of a kind was quickly restored.

Always the decisive, even impulsive, brother, Marcus had uncharacteristically made the Rothschilds wait for a commitment until Samuel, instinctively more cautious in business matters, had also been to Baku and both brothers had talked at length and in detail with Fred Lane. For they knew that once they had thrown their hats in the ring, there could be no turning back. They would be going to war with Standard Oil on a broad front and Rockefeller would take no prisoners.

The hazards, plainly, were great: everything they had worked for – the hard-won respect and trust, their current financial security and the sure prospect of a worry-free future – would be put at risk. But the potential profits were huge, of an entirely different order to anything Marcus had previously envisaged. For he was now confronting another of the industry's central truths, sharply defined by Occidental's chief Armand Hammer three-quarters of a century later: 'When you win in the oil business, you win big. When you lose, you lose plenty.'

A CLANDESTINE COUP

On his visit to Baku and the Black Sea port of Batum, Marcus had quickly divined that success or failure would depend primarily on transport and distribution.

The Nobels with *Zoroaster* and Standard with its Atlantic fleet had got things half right. Clearly, the key to transforming sound business into the prodigiously profitable lay in shipping kerosene in real volume rather than as case oil. But seen from a purely commercial perspective, the current vessels – even the largest – were still seriously flawed. They were too small, incapable in the main of carrying anything but oil and were, above all, accident-prone. No oil tanker presently afloat was permitted on safety grounds to transit the Suez Canal, a fact which condemned any vessel journeying between Europe and the Far East to sail round the Cape of Good Hope, adding thousands of miles to her voyage. Moreover, the great majority of such ships spent 50 per cent of their time running empty: having discharged their oil, their tanks were too foul to risk the loading of any other cargo even where this was physically possible. Keeping ships at sea on long voyages with empty holds had no appeal at all to a hard-nosed merchant like Marcus. But a worthwhile solution to the problem could not be a single new and radically different vessel. To create and retain a substantial market share, a fleet of tankers would be needed from the outset to ensure that customers were never obliged to look to a competitor for their kerosene.

Maintaining a continuity of supply meant that in addition to a fleet of ships, storage depots and tank facilities would also be needed throughout the Far East. For here was another oil industry instance of bigger being better. To attempt to enter the market on anything less than an international basis would be to invite Rockefeller to concentrate his awesome competitive firepower on small, easily identifiable and all too easily destroyed target areas. Marcus and Samuel thus dispatched their young nephews Mark and Joe Abrahams to find and develop storage-tank sites across Asia, and also to set up a distribution network using the agents and trading houses with whom the Samuels were now enjoying a second generation of business relationships.

There was another, crucial condition that had to be met if the venture was to succeed: everything about it had to be kept secret until launch day. Amazingly, given the numbers of people involved, in banks, trading houses, drawing offices, shipyards and government offices of one kind and another on two continents, it very nearly was.

But it was not all plain sailing. By the summer of 1891, almost a year before launch, there were press reports of anonymous but powerful Jewish merchants and financiers pressing for leave to take tankers through the Suez Canal. And then Russell & Arnholz, a high-profile firm of City solicitors, began a sustained parliamentary lobbying campaign – which included writing direct to Lord Salisbury, the Foreign Secretary – urging that such permission be denied on safety and what today would be called environmental grounds.

Citing the convention of client confidentiality, Russell & Arnholz declined to identify the party they were acting for, even when asked by

Salisbury whether they were representing a British interest. Salisbury's carefully phrased question and the lack of a meaningful answer were taken by many as confirmation that Rockefeller and Standard Oil were behind both the newspaper stories and the lobbying campaign.

Marcus, meanwhile, continued to batter away at the Canal authorities, pointing out that his vessels would be certified safe by the ultimate arbiters of all matters maritime, Lloyds of London. And the Rothschilds weighed in with their formidable political muscle too. It was, after all, the English Rothschilds who financed the purchase by Benjamin Disraeli of the shares that gave Britain control over the running of the Canal. The critically important permission was finally granted on 5 January 1892.

Game, set but not yet quite match to Marcus. For the first of his radical tankers had yet to be launched.

The vessels were the work of naval architect and consulting engineer James Fortescue Flannery, later to become better known as Sir Fortescue Flannery, Liberal and Coalition Unionist MP for a total of more than 20 years, representing constituencies in West Yorkshire and East Essex, president of the Institute of Marine Engineers, naval architect to the Crown Agents for the Colonies, knighted by Queen Victoria in 1899 and created a baron by King Edward VII in 1904.

Flannery, born on Merseyside in 1851, was yet another of those fussily bewhiskered, extravagantly moustachioed engineers with which Victorian Britain seemed abundantly endowed. But though he might in a poor light have looked like a refugee from the music hall stage, he was the genuine – and hugely talented – heavy industrial article.

He began his career as an office boy working for a Liverpool consulting engineer named David Campbell and went on to serve a five-year apprenticeship with the Britannia Engine Works in Birkenhead. In the evenings, Flannery attended night classes at the Liverpool School of Science, ultimately heading the examination lists for his year and winning the prestigious Derby Prize.

It was Flannery's breakthrough moment, because the prize was presented to him by Sir Edward Reed, a former chief constructor to the Royal Navy, who offered him a job in his London office. Flannery spent five years with Reed designing and inspecting machinery both at sea and ashore.

Flannery next went into business on his own account as a naval architect and consulting engineer, founding the firm – again redolent of the music hall – of Flannery, Baggally & Johnson. One of his first commissions came from Marcus Samuel and the result, the 5,010-ton *Murex*, was launched in West Hartlepool at 4 a.m. on Saturday, 28 May 1892. The vessel was described by the *Northern Daily Mail* as 'a splendid steel screw steamer' and *The Times* added that a fleet of similar vessels would soon follow for the transportation of petroleum in bulk through the Suez Canal.

The *Murex* was a technological triumph, of which Flannery was justifiably proud. She could carry 4,000 tons of kerosene quickly and with unparalleled safety in even the heaviest of weather. Flannery had greatly reduced the likelihood of explosion by designing tanks that permitted temperature-driven expansion and contraction of the oil cargo without leakage and without the excessive build-up of residual flammable gases when the tanks were emptied.

Bound for Batum, the *Murex* made her maiden voyage on 22 July 1892. On arrival in the Black Sea port, she loaded Bnito's kerosene and passed through the Suez Canal on 23 August heading for Freshwater Island, Singapore, and thence to the new depot established by Mark Abrahams in Bangkok.

And there was more and better to come. Flannery's groundbreaking design and layout meant that once the kerosene carried by the *Murex* – named like the rest of the fleet after an exotic shell in memory of Marcus Samuel Snr – had been discharged, her tanks could be speedily steam-cleaned to a standard that permitted food grains such as rice to be shipped home untainted. It was a massive bonus.

Thus, an East End Jew together with his brother and their young nephews, a Paris-based firm of international bankers, a burly shipping broker with a taste for bowler hats, highly polished boots and check-patterned coats, and a hirsute naval architect with political ambitions had combined to ambush an industry.

And what would ever afterwards be known as 'Marcus's coup' or 'the great coup' succeeded brilliantly. Meticulously planned and executed with panache and a remarkable sureness of touch, it left competitors such as the Nobels and Rockefeller reeling. Of sixty-nine oil cargoes to transit the Suez Canal by the end of 1895, only four belonged to anybody other than Marcus. His kerosene was on sale all over the Far East, a fact which totally disabled Rockefeller's finely honed – and for rivals, usually terminal – tactic of slashing prices in one market and underwriting the costs by raising them in another. He was not a happy tycoon.

A CRISIS CURED

Rockefeller could see only too clearly that Marcus and his allies had, at a stroke, effectively written the epitaph of the eastern case-oil trade from which he had been making massive profits.

But even as Standard's managerial representatives rushed about the Far East trying to quantify the damage to their Devoes brand – the name under which Rockefeller sold kerosene in the region – the one potentially disastrous flaw in Marcus's scheme was rapidly becoming apparent.

The coup had been based on the belief that success would be achieved by supplying kerosene of at least comparable quality to Standard's but, because of the major savings permitted by bulk shipping in the new tankers, at significantly lower cost to customers. And they would signify their approval of Shell Oil – it was called that from an early stage – by turning up in multitudes at distribution and sales points bringing their own cans with them.

Customers, however, were chiefly noticeable at this point by their absence. With full storage tanks at ports across the Far East and not a can-carrying punter in sight, something had clearly gone seriously wrong, a fact signalled by the forerunners of what would quickly become a flood of agitated cables to London. In truth, Marcus's great triumph was already in crisis and looked set to become a catastrophe.

What nobody in Houndsditch had fully appreciated was that the blue Devoes cans were valued by customers almost as highly as their contents. All over the Far East, the cans were reworked into roofing materials, door panels, cooking pots, containers of every conceivable kind, strainers, bug and rat-proof storage jars and a thousand other domestic utensils.

As soon as Marcus understood the problem, he produced a solution with the speed and ingenuity of the born 'can do' entrepreneur. Customers would of course be given tins if they wanted them. He chartered a ship, filled it with tinplate and told his regional partners to set up factories and workshops to make kerosene cans. That none had any experience to bring to the task was neither here nor there; everybody would be on the same steep learning curve but had only one thing to remember – that it *could* be done.

There were questions, sent by cable, on almost every aspect of manufacturing tin containers. All were swiftly dealt with – even if loosely and sometimes by way of little more than exhortation. From north China came a query about what colour the cans should be painted. Mark Abrahams responded saying that bright red would do very well.

It was an inspired choice because, whatever the aesthetics, it would now be possible to see who was winning the kerosene war in the region by watching roofs turning from Devoes blue to Shell red.

Bold action had thus cured the problem as quickly as it had arisen and by the end of 1893 another ten tankers – including the *Cowrie* and the *Conch* and *Clam* – had been launched. The business was booming and the Samuels were making serious money.

The good times were rolling now for Marcus; his campaigning in the Portsoken ward paid off and he was elected Alderman. Within three years, election to Sheriff of London would follow, leaving him only a single step away from the Lord Mayor's office he coveted.

Developments in the business were similarly encouraging. Marcus was paying Bnito for the oil he was transporting and selling in the Far East by way of bills of exchange at up to four months' sight after the cargo had been sold. He had similar arrangements with his major agents over the construction of key facilities such as storage tanks and wholesale and retail sales outlets.

The terms of trade were generous and were a tribute to the reputation – the 'good name' – of Samuel that Marcus and his father cared passionately about. The value of the work that father and son had invested in creating and maintaining their network of agents was now made manifest by the fact that each part of the new business was given time to generate income before bills fell due.

It also meant that at any one time, the Samuels were handling a great many bills of exchange and, in fact, acting as merchant bankers. Marcus and Samuel thus decided to do the job properly by setting up and becoming co-owners of a private bank that in time would become part of Hill Samuel.

Just when it seemed that everything was going extraordinarily well, Marcus was given profoundly shocking news. He had become seriously unwell and shortly before his 40th birthday was told by his doctor that he had cancer. The prognosis was dire: he could not expect to live more than a few months, certainly less than a year.

Both diagnosis and prognosis were, of course, to prove utterly wrong and Marcus would live for another 34 years. But beyond the initial fearful impact, the threat of imminent death made him focus his attention on bringing some sort of order into his own tangled affairs. For the truth was that as an entrepreneur, an originator of ideas, a spotter of opportunities, a converter of half chances and an intuitive deal-maker, Marcus was peerless. As the director of a major business that had necessarily begun life as a multinational touching the daily lives of millions of people across the world, he was a disorganised disaster waiting to happen. Marcus loved doing business; administration, organisation and the writing down of detail was stuff that got in the way. He had been flying on flair and by the seat of his pants for too long, and he knew it.

One of the consequences of being forced, even if temporarily, down to earth by intimations of mortality was the formation of the Tank Syndicate.

Nothing like as secretive or sinister as its name suggested, the Tank Syndicate was the true parent of Shell and comprised the two Samuel brothers, the shipping broker Fred Lane and the seven Far Eastern agents involved in the oil business. Together, they agreed that profits, losses and liabilities would be shared. Collectively, they believed they could muster the strength to fight Rockefeller should he choose to attack in any of their markets.

But also on Marcus's mind was the security – or otherwise – of his oil supply from Bnito. For, as events almost exactly 110 years later would amply demonstrate, an oil company with no oil – or even sharply diminished reserves – is as much an object of derision as a pub with no beer.

The problem with Bnito came in three parts. First, there was the crazily unpredictable price the Rothschild company charged for transporting oil overland between Baku and Batum. The knock-on effects for buyers and shippers were as obvious as they were unwelcome. There was also the deeply troubling matter of Marcus's total dependence on Bnito for oil. He owned neither wells nor production companies. When the original ten-year contract with Bnito expired – and it was already at the halfway point – he would be in an exposed and vulnerable position, obliged to pay whatever Bnito might demand just to stay in business.

And looming over everything was the political volatility of the region. The Tsarist government had a disturbing history of oscillating violently between contradictory positions. Even the existence of Baku as an oil-exporting entity was subject to the Tsar's whim. And Tsars had in recent times proved themselves capable of both extreme xenophobia – the Nobels' father Emmanuel had been obliged to watch the dismemberment of the Russian industrial empire he had patiently constructed when the Tsar suddenly took against foreign ownership of businesses – and crude anti-Semitism. To be sure, the Rothschilds seemed immune, but then many in Russian business believed the Rothschilds could walk on water.

There was much to lose because the Tank Syndicate was thriving. Moreover, the Samuels were also making money from more than oil. With Britain adopting a stance of neutrality, the brothers became major arms suppliers to both sides in the Sino-Japanese war of 1894–5. Japan's victory produced a substantial bonus for them, too, in the form of new contracts which consolidated the status of Samuel Samuel & Co. as the leading British company trading in and with Japan.

Palpably not dying of cancer, Marcus was now thoroughly enjoying his wealth. He drove each day from Portland Place to the office in Houndsditch in his own carriage and he also rode every morning in Hyde Park. Witnesses casting an informed eye over his equestrianism said Marcus offered ample proof of the proposition that a horse was an animal dangerous at both ends and uncomfortable in the middle. But style in the saddle mattered much less than the fact of keeping and riding horses for pleasure in the park. You had to be very rich indeed to be able to do that.

Now an alderman and a familiar and respected figure in frock coat and top hat in the City, Marcus had developed a touch of pomposity to go with his fierce patriotism. He once upbraided a clerk who announced him as 'Mr Marcus Samuel' when he arrived at the office of a friend. 'I am,' said Marcus, standing firmly on his dignity, 'Mr Alderman Samuel.' Recognition, status, acceptance had been too hard won to be dismissed lightly now.

But concern over the Tank Syndicate's total reliance on Bnito for oil would not go away. Like Rockefeller with Standard, the Syndicate simply had to have its own oil and, moreover, control over every stage involved in its production, refining, transport, distribution and sale. Mark and Joe Abrahams, having completed their task of setting up storage and distribution facilities in the Far East, were dispatched to find oil.

Borneo was the chosen search area partly because oil was known to exist there but also because Marcus, when travelling through the region, had met a Dutchman with a 2,500-square mile concession in Kutei. Jacobus Menten had been working his concession for years without making a strike; he had, however, found oil seeping into streams. But this, according to a geologist who had analysed the samples, had a consistency like that of treacle in a cold climate, making it unsuitable for refining into kerosene.

Oil had also been found – and, indeed, had been known to exist for more than 400 years – in nearby Sumatra, where several exploration companies, including the memorably named Royal Dutch Company for the Exploitation of Petroleum Wells in the Netherlands Indies, were in need of support. Menten, however, managed to prevail on Marcus to the extent of receiving £1,200 to continue his search and £2,400 in royalties should he find oil.

With the benefit of a bare two weeks' training in oil exploration and production techniques, Mark Abrahams joined Menten in the jungle in October 1896 and, having cleared a site in Sanga Sanga about 50 miles from Kutei, work began.

After drilling through sandstone to a depth of 150 feet, on 5 February 1897 Mark struck oil. That was the good news. The bad news was that the oil was remarkably similar to that found by Menten and sent for analysis earlier. It, too, would be hard and expensive to refine into kerosene. To rub salt into the wound, a gusher was struck in a concession which Marcus could have bought a few miles away. The oil roaring out of that well was a light crude perfect for conversion into kerosene. But Marcus remained extraordinarily bullish about the Kutei find, predicting that it would produce between 15 and 20 million cases.

It was partly because of this that talks aimed at a full union, a real amalgamation, between the Syndicate and the Royal Dutch Petroleum Company, the third party at work in the area, were stopped dead in their tracks.

The main reason for the halt, however, was the implacable opposition to any transfer of Royal Dutch assets by the company's young and formidably ambitious marketing director. His name was Henri Deterding.

CHAPTER 2
THE UPPITY YOUNGSTER

On 19 April 1866, the wife of a captain in the Dutch merchant marine gave birth at home in Amsterdam to a healthy boy. The fourth of five children, the boy was christened Hendrik August Wilhelm but would become widely known in the industrialised world as Sir Henri Deterding and, as the autocratic and megalomaniacal boss of Royal Dutch/Shell, revered and feared in roughly equal measure.

Deterding had two older brothers whose education and careers – one was to become a doctor, the other an army officer – had been meticulously planned for them by their solidly middle-class parents. Young Henri, by contrast, wanted only to follow in his father's footsteps. He would say later that he became entranced by the sea in early childhood and spent every moment of his free time on quaysides in the company of ships and seafarers.

Disaster struck early in Deterding's life when his father, at the age of 42, contracted a fever and died during a Far Eastern voyage. Henri, described as 'a healthy boy full of the zest for life', was just six years old when the family's descent into genteel, but painfully real, poverty began.

There are many myths and legends – some of which Deterding himself helped manufacture and propagate – about the early life and times of a man who would be described by some contemporary newspaper commentators as amongst the most powerful in the world. That he received little or no formal education is one of the more persistent.

On the contrary, Deterding, according to Dutch state school reports examined by biographer Glyn Roberts, proved to be 'a good boy, a sound, healthy, normal pupil' and, while still a little under ten years old, 'competent in the Dutch manner, serious and systematic in his studies'. And what was Deterding learning? 'Holland's history, her function as a nation, how she triumphed over apparently impossible odds, how she lived under the permanent threat of destruction by the sea.' Most of all, however, Deterding was learning exactly what had made Holland such a prosperous merchant trading nation having already sensed 'the phenomenal respect of the average Dutch middle-class family for private property and its protection'. Crucially, the young Deterding had determined at his desk to make a place for himself in Dutch commerce.

Precisely how this was to be achieved was very much open to question. For while he studied, among much else, French, German and English in addition to his native Dutch, Deterding had already been told by his mother that there would not be enough money for him to continue his education beyond the age of 16.

Recalling that conversation many years later, Deterding said that it had given him 'an almost adult vision. Though a boy in years, I ceased to have a boy's outlook. I began to think like a man and thenceforward, I resolved my career must be in business.'

Deterding thus left school at 16 and started work with the Dutch Twentsche Bank in Amsterdam. He was their youngest employee and quickly discovered that he disliked both banking and what he regarded as heavy-hoofed, pedestrian bankers. For their part, the honest plodders of De Twentsche made it abundantly plain that they didn't much care for the pushy, blundering Deterding, a young man in altogether too much of a hurry for the sedate traditions of the counting house.

In truth, the bank kept young Deterding's nose within a whisker of the grindstone most of the time and, boring and laborious though he found much of the work, he came later to regard this period as time well spent. It was, he believed, excellent experience which yielded many insights into the foibles and weaknesses of men as well as their strengths and qualities.

By and by, Deterding was set to work under a new departmental head, a man named Vanderbyll, who appeared to have much more time for the uppity youngster than the bank's other senior staff. Vanderbyll had strong opinions on many matters and not least the problems of finance. In Deterding he found a sharply intelligent and appreciative audience, especially when he permitted the junior man to listen in while he dispensed advice to those of the bank's clients with substantial capital to invest.

Deterding developed a profound respect for Vanderbyll and many years later, when he was at the zenith of his career and power, continued to correspond with him. In fact, a regular exchange of letters went on well into Deterding's retirement.

The probability is that Vanderbyll was among the first to recognise Deterding's passion – and it was nothing less – for figures and his extraordinary ability to decode, literally at a glance, the true and underlying state of a company's financial health from its balance sheet. Almost everybody in the bank knew by now that Deterding also boasted another unusual gift in the form of an astonishing facility in mental arithmetic, something he would coincidentally share with John D. Rockefeller and, of a later vintage, Mobil's Bill Tavoulareas, widely regarded as the ablest oilman of his generation.

Deterding himself was powerfully aware that his talents – esoteric as they were – not only set him apart from his colleagues but would also, in future, give him a formidable advantage over business competitors. He took great care to keep his skills honed and polished by making the study of figures and balance sheets his all-consuming hobby, a fact which established Deterding as a 24-carat oddity even in a business replete with characters whose eccentricities could easily be mistaken for barking madness.

Under Vanderbyll's watchful eye, Deterding was transferred to the bank's stock department, where he quickly absorbed and mastered the arcana of share trading, company acquisitions, asset transfers and investment procedures. And this knowledge, coupled with his phenomenal ability to make lightning calculations having already unravelled a bunch of balance sheets, equipped him to become the man whose 'middle name' – according to contemporary American financial writers and business journalists – 'was merger'. Deterding earned the tag by virtue of the number and complexity of the corporate marriages and amalgamations he forced through in his long and increasingly controversial career.

After six years of solid, if not always universally appreciated, work at De Twentsche Bank, Deterding became dissatisfied and convinced that he was worth considerably more than he was being paid. He was determined to move on and began scanning newspaper advertisements for 'interesting business openings' and also occasionally sat examinations in the hope of securing 'something better'.

His breakthrough came when – as a result of the single-minded application that would become his trademark – he topped the list of candidates competing for a position in the colonial organisation of Nederlandsche Handelsmaatschappij (Netherlands Trading Society), the Dutch equivalent in prestige and history of London's East India Company. The Society was in reality a banking organisation interested in almost every aspect of Dutch trade, and it worked closely with both the government and the country's royal family. The Society's primary task was ensuring that a broad stream of dividends flowed into Amsterdam not only from Holland's dominions but also from a variety of other sources.

Deterding, 'half happy, half melancholy', set off for the East Indies wondering where he might be in a few years' time and also, crucially, when he would be able to manoeuvre himself into a position in which he would be able to reap the profits of his work for himself.

Given that he had rarely gone beyond the confines of Amsterdam in his early life, Deterding was to experience initial loneliness and no small degree of awkwardness on arrival in the East; he had, after all, no friends or relations in the territory and little or no knowledge of the customs and conventions of Dutch colonial society. He could be sure only of the fact that his enthusiasm for the instant analysis of balance sheets would be shared by few of his fellow expatriates.

Within a month of arrival, however, he was joined by C.J.K. Van Aalst, a young Dutchman he knew well and the man, moreover, he had recently beaten into second place in the Society's examinations. The two were to maintain close business contact for almost four decades. And both men would in mid-career be honoured with knighthoods in Britain in recognition of their wartime service – a fact which in the case of Sir Henri would later come to be regarded as a gruesome irony. For by then Deterding had achieved a curious and mostly unwanted celebrity at the head of what was the second biggest corporation on earth, an organisation of prodigious power and wealth. Van Aalst, while also a director on Deterding's board, had meanwhile become the Society's chairman and also an influential international financier in his own right.

NIGHTMARE IN DELI

In the East, Deterding was given an unexciting but extraordinarily exacting start by the Society. He was dispatched to a small and utterly remote town called Deli. And if it had not already existed, Conrad would surely have invented Deli as precisely the sort of place in which one of his characters, after a period of fevered introspection, would disintegrate into sweat-soaked madness.

Deli's claim on Deterding's undivided attention and formidable powers of concentration stemmed from two facts: there was a Society branch office there and, in the office, a set of books in an unspeakable mess.

When first confronted by the chaos, Deterding – one of nature's accountants, to be sure, and a man seriously in love with numbers – said he did not know where or even how to begin. After a long and characteristically thorough examination of the ledgers, he was of the opinion that bringing some sort of order to bear might actually prove impossible. The books, it seemed, had been balanced monthly but never correctly. Thus, the incorrect figures for any month you cared to examine were, in part at least, the compounded product of the previous month's errors and so on.

Here, on the edge of the jungle in an office where a fan slowly churned the thick, wet, superheated air, was a numerical nightmare. Here, in Deli, you can sense disbelief turning into panic as the realisation strikes that there's simply no way of knowing how much of each month's crop of errors is historic and how much is new, nor is there any clue to when, exactly, the first calamitous miscalculation was made. A sane man would of course have closed the ledgers, had a stiff drink and packed his bags. Deterding, however, had this to say of the episode:

> Had I let that tough bookkeeping job at Deli master me . . . my whole life would have been different. A minor bookkeeper I then was, and a minor bookkeeper I should probably still be had I not grappled with that confused mess of figures. The supreme advantage of unravelling those figures was that it gave me the unravelling habit.

The way Deterding tackled the task was revealing of his character. Having made an assessment of the magnitude of the problem presented by the delinquent Deli books on a Friday evening, Deterding left the office and spent the entire weekend walking and thinking.

On Monday morning and with the ledgers spread before him, he believed he could see a way forward. But it was not until Thursday that he was sure he had the key to unlocking the whole convoluted business. He began by making a forensic scrutiny of the business records of the past 18 months while, simultaneously, creating new accounts for every transaction – however minor – made on each and every day of the past year and a half. There were thousands. And he was, of course, obliged to work backwards.

It was an immense, almost Sisyphean labour even for somebody who, like Deterding, was obsessed with figures and driven by a pathological determination to succeed. There was no real choice: he had indeed to 'master' the problem, make it succumb to his will. For Deterding knew that this preposterous task presented him with a brilliant shop window for his strange blend of talents, appetites and skills. Here was a chance to shine so brightly that the big bosses of the Society would be compelled to take notice of the recent arrival from Amsterdam, a small, neat man with unusually bright, laser-like eyes.

Deterding succeeded splendidly. Four months after embarking on the deeply flawed Deli books, he brought solid Dutch order out of chaos. And beyond the notice he craved, he was rewarded with a thumping 75 per cent salary increase – the biggest, he would subsequently always claim, he ever received. Best of all, despite his youthful looks, Deterding was asked to act as the Society's temporary agent in Medan.

This was a major and critically important promotion for Deterding. Medan was among the most commercially significant entrepôts in the Dutch East Indies and a bustling centre of a huge array of trades and businesses. Moreover, Medan was also home to a colourful, polyglot army of adventurers and chancers who spent much time and energy searching for loans.

The Society had been trading in Medan since 1824 but by the 1880s its chief activity was banking and the active management of interests in rubber, tea and tobacco plantations, as well as important investments in a wide range of other industries and enterprises. Deterding, with by now customary single-mindedness, set about mastering his new responsibilities. To be effective in his new post, he had to learn – and in short order – something at least about the score or more of trades most important to the Society in Medan. This meant keeping himself informed on a daily basis about the multitude of factors, some technical, others regional, all of them specific, affecting those businesses and thus the health of the Society's loans. He soon discovered the importance of being able to make almost instant decisions about the character of the men appearing in his office in pursuit of loans. But he also quickly realised that he needed, wherever possible, a sound independent source of information about the reliability of these potential clients. He thus developed another talent for what he called 'sniftering around'. Although he struggled to define exactly what he meant by 'sniftering' – his difficulties might well have stemmed from the need to be defensive about activities that many of the Society's clients would have found offensive – it appears to have amounted to the establishment of an elaborate intelligence jigsaw. Deterding had discovered for himself the primacy of information in business generally and in the business of making money in particular.

'Without this flair for sniftering,' Deterding was later to declare, 'no man starting from the bottom can make money on a large scale.' And the making of money on the largest possible scale was now Deterding's sole aim and chief ambition. Every weapon in his arsenal – a talent for sniftering, the love of figures, the speed-reading of balance sheets, the ability to make lightning calculations without recourse to pen, paper or abacus, the swift judgement of a man's character – was deployed in its realisation.

And Deterding did, indeed, make money. Branch turnover soared as he found new business and far more profitable ways of handling existing trade. He also began to focus on the movements of discount rates in the world's major money markets and learned, within reasonable limits, to predict both the scale and timing of fluctuations. This was serious business and it began to produce serious profits.

Deterding was, in late 1894, transferred from Medan to Penang and had scarcely unpacked his bags when he seized on the Telegraphic Transfer, that newest of technological aids to banking, to conduct a large and lively trade in the buying and selling of bills. He was soon making big money. The problem, from his viewpoint, was that the chief and sole beneficiary was the Society.

He began, not unreasonably, to ask for more and, never a man to sell himself short, also made it plain to the Society's top brass that they were greatly blessed to have his services at such a bargain rate. Deterding was taken aback when the Society responded with the view that their man in Penang was not entirely without faults. His official reports and correspondence, for instance, were often flawed by loose writing and a shocking disregard for stylistic niceties. Deterding, thoroughly angry, replied, saying that he believed himself employed to make money rather than produce literary masterpieces. In truth, his relations with the Society were in tatters when, perhaps fortuitously, he fell ill with a throat infection.

Deterding applied for time off to consult a good doctor and the Society, acknowledging his valuable work, awarded him two months' medical leave in the more salubrious surroundings of Magelang. Deterding's temper cooled. It was to boil over again within days, however, for he had barely arrived in Magelang when he was summoned back to Penang because the Society had caught a Chinese cashier falsifying accounts. Deterding was ordered to return immediately to investigate the matter.

Still smarting from what he considered to be unmerited ill treatment as well as the Society's outrageous parsimony in denying him a share of the Penang profits, Deterding took up pen and paper and wrote two letters.

Addressing himself to the Society, he pointed out that he had warned his superiors several times about the conduct of the crooked cashier. The whole distasteful affair could have been avoided had they taken notice – and action – earlier. He went on: 'I would rather not work for the Netherlands Trading Society if I am never to share in the profits of Penang but am called back, notwithstanding, when something goes wrong in my absence.'

The second letter was addressed to J.B.A. Kessler, who six months earlier and several times in between had made him the offer of a job. Deterding told him he was now ready to accept. It was a momentous decision because Jean Baptiste August Kessler was running an oil company called Royal Dutch.

WELLS RUN DRY

Deterding's opposition to an amalgamation of Royal Dutch and the Tank Syndicate came at a time of intense, intricate manoeuvring not only by the two principal companies but also by Standard Oil.

As Rockefeller was by now old and infirm, Standard was under the control of John D. Archbold. Of Irish descent and mercurial, high-spirited, jokey, decisive and with charm in abundance, Archbold was worshipped by his staff at 26 Broadway and was the antithesis of the increasingly arthritic *éminence grise* who had appointed him.

But, like Rockefeller, Archbold was a controversial figure and no stranger to the courtroom and witness box. He had been roundly condemned for his 'extreme evasiveness' while giving evidence during the investigation of the Standard Oil Trust and had also been accused of bribery in another case.

Beneath the easy charm and the even easier smile, Archbold was a case-hardened oil industry professional who combined aggression with flexibility. Like Rockefeller, all of whose works and interests he had once forcefully and publicly denounced at the time of the railway freight scandal, Archbold had been in the thick of the action from the early days. A veteran of the Pennsylvania oil region, he had served as secretary of the Titusville Oil Exchange and knew almost everybody who mattered. And he was the only American oil industry figure of consequence to insist back in 1893 that Marcus Samuel should be paid serious attention.

In 1897, Archbold instructed W.H. Libby, Standard's ambassador-at-large, to engage both Marcus and the Rothschilds in talks and also open negotiations with Kessler and Royal Dutch. For Archbold and Standard were keenly aware that American oilfields were finite and were anxious to establish footholds in Russian and Far Eastern production.

Superficially, Royal Dutch looked to be in a strong position. Under the able and energetic leadership of Kessler, who in often appalling jungle conditions and at huge cost to his health had pulled the company back from the edge of extinction, Royal Dutch now had oil in quantity. But though Kessler was about to launch Crown Oil, a Royal Dutch brand of kerosene refined in-house, the company's sales and distribution networks were frail and its marketing feeble.

The Tank Syndicate, by contrast, had superb transport, distribution and sales networks but was still reliant on its Bnito contract and its single Kutei source for oil. In the circumstances, which included the looming presence of Archbold and the predatory Standard in the shadows, an amalgamation had much to recommend it.

Marcus, however, was sanguine in the face of Deterding's opposition.

He and Kessler had time and much respect for each other, and agreed that the way should be kept open for a possible combination in future. But Marcus and Sam Samuel also had many other pressing demands on their attention.

First, there was the move of office from the old Houndsditch premises to the larger and more impressive accommodation they had found in Leadenhall Street. Once again, the distance involved was small – little more than a mile – but in terms of prestige the difference was huge.

The business of the move was, of course, turned into theatre – farce, in fact – as removal vans were obliged to remain motionless in the narrow streets, blocking traffic in all directions, while the brothers huffed and puffed over documents. First the Alderman refused to give up files he claimed to be reading only to be followed by brother Samuel declining to release accounts he said were in need of scrutiny there and then. By and by, both men and their filing cabinets were relocated so that the traffic could flow once more.

The brothers celebrated their move with nothing less than another, though very different, coup – the floating of the first Japanese government sterling loan in the City of London in 1896. In the amount of £4.5 million, a huge sum at the time, the loan, as presented by the Samuels to the City, was oversubscribed in a matter of hours. It was a triumph and underscored the unique relationship of trust the brothers had fostered in Japan; other, similar transactions would follow in the years ahead, most notably for infrastructure work and a major railway project in the Osaka and Yokohama areas.

And Marcus's civic career was flourishing too. Having completed his term as Sheriff of the City of London, he was appointed one of the two aldermen who heard cases of appeal in the County of London Sessions and was also made chairman of a special committee giving evidence to a Royal Commission on the Port of London. At the same time, he was given the news that he would be made a Justice of the Peace in Kent and he was also elected Master of the Spectacle Makers' Company, one of the City's most ancient Guilds.

Marcus was, in truth, well on his way to becoming a public figure and although there were clear and present benefits for the business, the mushrooming demands on his time were also taking him away from it more often and for longer.

But there was one matter to which he gave his full attention: the formal launch, after much detailed devilling by lawyers, on 18 October 1897 of the Shell Transport and Trading Company Limited. The original capital was £1.8 million in ordinary shares with a nominal value of £100 each. Of these, Marcus took 7,500 shares and Samuel 4,500, contributing between them £1.2 million representing the full value of the Shell fleet and some of the Kutei assets. Other members of the Tank Syndicate received shares in proportion to their expenditure on tank installations and other assets they had contributed.

The shares held by Marcus and Samuel, however, had special voting rights of 5:1 over all other ordinary shares, ensuring that Marcus retained complete control of the company. He would not surrender this for more than 20 years, in fact until 1919 when he agreed to waive this right at the specific request of the British government.

If 1897 was closing on a high note for the Samuels and their new Shell company, it would end in disaster for Royal Dutch. For on 31 December, Kessler's wells began to produce an oil-and-water mixture, an unfailing sign that they were close to exhaustion. Within three months, most of the wells were dry.

Production from Shell's Sanga Sanga wells had fallen to just 15 bpd when on 15 April 1898, Mark Abrahams again struck oil, at 750 feet, with the first well he drilled at a new site at Balik Papan. It was an important find, as it immediately boosted production to 130 bpd, the equivalent of 200,000 cases per year.

And fortune had also smiled on Marcus in another and altogether extraordinary way. On 15 February, the British warship HMS *Victorious* had run aground at the entrance to the Suez Canal. It was an incident which caused considerable embarrassment in Britain and much speculation overseas, in no way lessened by the *Daily Mail* on 16 February suggesting that the whole thing was a ploy by the Admiralty 'to prevent warships of other nations from making a quick passage to the East'.

The Lords of the Admiralty, seeking to prevent an undignified drama turning into a crisis, immediately ordered the cruisers *Gibraltar* and *Hawke* to give assistance. They were of no help. Nor were the tugs of the Suez Canal Company, a P&O steamer and sundry other vessels. When Marcus read that the warship was still stuck fast, he cabled Worms et Cie, Shell's agent on the Canal, to order *Pectan*, the world's most powerful tug and just commissioned by Shell, to offer assistance. She did, it was accepted and in a 21-hour operation, HMS *Victorious* was dragged clear.

The *Pectan*'s crew shared a reward of £500 put up by a grateful Admiralty who paid her master a similar amount. They offered her owners £5,000, a sum Marcus waved away, saying that, as a patriotic British citizen, he had been only too pleased to help. He did not, however, dismiss the knighthood he was offered instead and, on 6 August, went to Osborne House, the royal residence on the Isle of Wight, where Queen Victoria made him Sir Marcus Samuel at a private investiture.

It was a moment to savour and cherish because in business the good times were all but over for Marcus.

A SEA OF TROUBLES

Following the failure of its wells, Royal Dutch was also obliged to buy Russian kerosene and began to sell it in competition with Shell in the Far East. Kessler, painfully aware of the weakness of his company – Royal Dutch shares had crashed on the Amsterdam bourse within hours of the bad news being telegraphed to Europe – was fearful that Marcus and Shell would now launch a Standard-style price war he could not possibly survive.

Curiously, both Marcus and Archbold missed the opportunity of buying Royal Dutch shares, and even a controlling interest, at a bargain-basement price. Instead, Kessler was surprised and relieved when Marcus, observing that the regional market was big enough for both of them, offered an agreement – in reality, a defence pact against Standard – under which Royal Dutch and Shell undertook not to undercut each other in the Far East.

Kessler, of course, was delighted to sign. But Marcus was also pleased to have made the deal because he had other, bigger matters to resolve.

First was the question of securing a new oil supply given that his contract with Bnito would expire in late 1900. He came to terms with Moeara Enim – a company Archbold had only narrowly failed to acquire for Standard – and a contract was signed specifying that Shell would handle all Moeara Enim's exports of illumination-quality oil. However, because the amounts were relatively small, a cancellation clause was written in permitting termination of the agreement if the minimum annual supply had not reached 50,000 tons by 1 January 1901.

With that deal concluded, Marcus embarked on a subject that would occupy him for the next 15 years. He believed that rather than incurring the costs of turning his heavy Kutei oil into kerosene, it could be used to fuel Shell's own fleet of tankers. The Nobels, after all, had shown with the *Zoroaster* that oil-firing was feasible.

And if commercial vessels could be run on liquid fuel, why not those of the Royal Navy? It was a huge ambition because Britain's fleet was easily the biggest in the world; and it was an ambition that would cause Marcus pain, humiliation and great loss before, finally, the idea was turned into reality.

It was former British Prime Minister Harold Macmillan who, when asked by a reporter what he had feared most during his premiership, said: 'Events, dear boy, events . . .' It was a sentiment Sir Marcus Samuel would most certainly have endorsed at the beginning of the twentieth century.

The oil business was changing beyond recognition. Always large in scale and international in both its impact and dealings from the earliest days, the industry had quickly become extraordinarily complex. Driven still by improbably vast profits, it was now a highly politicised business, full of uncertain loyalties and shifting alliances, punctuated by sudden, savage battles and tortured truces, and orchestrated by crunching conflicts of interest. Innocence, and much else besides, had been lost in the profits of Pithole.

And Marcus Samuel had changed too. Now in his mid-40s, he had aged rapidly and far from gracefully. Always 'stout' – a quintessentially Victorian euphemism he employed to describe both himself and his even more overweight brother Sam – he was now a short, fat, middle-aged man with much grey in his hair. The soft-spoken confidence, immense vitality and optimistic eagerness that had hallmarked his early career, engendering trust and fuelling friendships, had turned into pomposity and a tendency towards preachy sententiousness. He did not like having his judgements questioned and he rarely asked for, and was anyway not much interested in, the opinions of anybody else.

The process of withdrawal that close associates had noted earlier had not only continued but also, if anything, accelerated. He often retreated now behind a watchful inscrutability, his face mask-like. Where once there had been a degree of warmth and shy, engaging charm, there was now coldness. For the truth was that although he had been an alderman for half a dozen years, Sheriff of London and knighted by Queen Victoria, he felt himself still – and perhaps even more – an outsider denied his just measure of recognition.

He had, to be sure, made an immense fortune – in reality, several distinct fortunes – and with a grand London house and a vast country estate, he had all the trappings of great wealth and the externals of tycoonery. Moreover, he had been generous in giving to the charities and causes he thought worthy of support, and, crucially, had made his money honestly. In an industry in which bribery was common currency, Marcus was never once accused of buying influence, favours or preferment.

Social success had, however, despite his titles and honours, eluded him. His Jewishness set him apart, but his co-religionists in the Anglo-Jewish aristocracy had shown themselves unenthusiastic both about receiving him in their homes and accepting his offers of hospitality.

Marcus was lonely and, with Sam away in Borneo for almost nine months investigating baseless allegations laid by Jacobus Menten about Mark Abraham's conduct, in sole charge of Shell at a particularly dangerous time.

Standard, in the ceaseless corporate manoeuvring of the time, had approached both the Nobels and the Rothschilds with a proposal for a grand alliance. Marcus's response was for Shell to lay down a barrage of activity on all fronts in a demonstration of his company's might and muscle.

He quickly made moves to foster the relationship of mutual interest he had formed with Vice-Admiral Sir John Fisher – the 'oil maniac' – who was almost certainly the only other man in England to share Marcus's evangelistic enthusiasm for oil-fuelling the fleet. And having in June 1899 attended the first public trials of horseless carriages arranged by the Royal Automobile Club and decided, correctly, that motoring was going to be the next big thing, Marcus instructed Mark Abrahams to set up a petrol refinery in Balik Papan. He was determined that Shell would be at the forefront of supplying fuel for the motors.

Next came the expansion of the Shell fleet. Marcus was already building two new 9,000-ton tankers, the biggest in the world. Now he laid down a third and forced a fourth onto Fortescue Flannery's drawing board. And he began buying oil in vast quantities where and whenever he could. The market was strong – particularly in the Far East where, as he had predicted, demand was racing ahead of supply – and prices were rising. But the risks were similarly huge. When Shell's tanker capacity proved inadequate, Marcus chartered to ensure his company could handle all the oil he was buying. At one time Shell had no fewer than 16 tankers in the Suez Canal.

Sam, on his return to Leadenhall Street just before Christmas, was horrified. He and Marcus had the biggest, most explosive row of their lives, in which each called the other fool, lunatic and imbecile. Marcus's risk-taking and spending of money like a drunken sailor had appalled and frightened the instinctively more cautious Sam. Marcus's whispered reply was for Sam to look at the balance sheet.

And at the June 1900 AGM, the balance sheet looked remarkably good. Profits were up by 60 per cent and at a level of more than £1,000 per day. The trading value of the ordinary £100 shares had tripled, and the dividend for 1899 would be 8 per cent, with a 5 per cent dividend for the first half of 1900. Marcus also proposed a series of technical changes to ease the buying and selling of Shell shares, including a warrant system for the Far East, and a £200,000 increase in the company's capital. The 252 delighted shareholders agreed readily to everything proposed. The show of muscle Marcus had staged also had its desired effect: Standard's proposed alliance with the Nobels and Rothschilds came to nothing. Marcus was thus able to renegotiate the Bnito contract without difficulty. There was euphoria in the air. But it was to turn to crisis almost immediately as a huge, destructive wave of events engulfed Marcus and Shell.

First came a collapse in the price of kerosene just as Shell had stockpiled immense quantities. It could be sold now only at a loss. In China, the company's storage tanks and other installations in Canton, Hangchow, Tientsin and Shanghai were looted and severely damaged during the Boxer Rebellion. In South Africa, the Boer War between the British government and Dutch settlers had effectively halted a new marketing venture. Worse, the ferocity and bitterness of the 'black week' fighting at Colenso, Stormberg and Magersfontein caused an inevitable upsurge in Dutch nationalism worldwide. In the Far East, there were tensions at some sites between Dutch and British officials and staff, and at one stage the Dutch government ruled that only nationally flagged vessels would be permitted to trade with the East Indies colonies.

The ruling was made at a time when Henri Deterding was pushing hard to increase the sales of Royal Dutch in the region. He had set up his own agents and installations in at least five ports in direct and increasingly sharp competition with Shell, even though both companies were selling Russian-sourced kerosene. The price-fixing arrangements brokered by Marcus Samuel and Jean Baptiste Kessler had not been abrogated; rather they had been allowed to lapse.

The news from India was no better. Marcus had made substantial investments in the subcontinent only to see them nullified when Burmah Oil won control of the kerosene market. And things were little better on the home front because Standard, despite Marcus's enthusiasm for horseless carriages, had stolen a march on Shell and sewn up the majority of 'motor spirit' – petrol – outlets.

On top of everything else came a general recession and a major downturn in world trade, with a consequent sharp fall in shipping freight rates. Shell's splendid fleet of tankers, so artfully fashioned by Fortescue Flannery to be capable of accepting all kinds of cargoes in steam-cleaned holds, could now only be operated at a loss in the new conditions.

Just when it seemed that nothing more could conceivably happen to worsen Shell's position there was cabled news, a little before Christmas, that Kessler had died.

More than any other individual, Kessler had been responsible for the survival of Royal Dutch when, twice at least, the company had been within an ace of terminal collapse. He had led from the front, sparing no effort, and by personal example. The cost to his health – never robust – had been huge.

Following the calamitous failure of its wells, Royal Dutch, urged on by Kessler, had in the late 1890s embarked on a drastic programme of exploration. The company had drilled no fewer than 110 dry holes in Sumatra in a bid to find new oil. About 80 miles north of its existing Sumatran concession, in a remote frontier area called Perlak, the company found a seepage. Hugo Loudon, a young engineer of great ability who boasted an unusual breadth of experience, led an expedition to the area. The son of a former governor general of the East Indies, Loudon – a future chairman of the company – had useful diplomatic skills, which, given that Perlak was still subject to a native rebellion, would be tested to the full.

Loudon negotiated with both the Rajah of Perlak and the rebel chief, and arrangements satisfactory to all parties were eventually concluded. There were several geologists in his party and an exploratory well site was swiftly located. Drilling began on 22 December and oil was struck six days later.

Now, 11 months on and with Royal Dutch very much back in business to the tune of 5,000 bpd from Perlak, Kessler, worn out and racked with fever, cabled the company's offices in The Hague saying he was in 'a very nervous condition' and heading home. He had reached Naples when, on 14 December 1900, he died of a heart attack.

Shell's amicable relations with Royal Dutch depended to a large extent on the regard, based on trust and respect, Jean Baptiste Kessler and Marcus Samuel had for each other. Within 24 hours of Kessler's death, Royal Dutch appointed Henri Deterding to succeed him. The move would herald a new and far more aggressive era for Royal Dutch. For Marcus, Deterding's appointment would prove fatal.

CHAPTER 3

THE MOTHER GUSHER

Deterding, just 34, had a dream start in charge of Royal Dutch. Production from the Perlak wells was going from strength to strength at a steady, consistent rate. The company was back in the big game.

For Marcus, by contrast, the future was looking decidedly bleak. On top of all his many other troubles, the Moeara Enim oil source proved a cruel let-down. After showing early promise, the company's wells had run dry. In a letter to Marcus, they said they would be unable to meet their obligations; it was not a question of being unable to supply an annual minimum of 50,000 tons, rather it was a matter of not being able to supply any oil at all for two years.

But then, quite suddenly, a new source came blasting and roaring out of the ground on the other side of the world. Marcus had been thrown a lifeline by yet another of the extraordinary obsessives with which oil history is replete.

Patillo Higgins was a tall, burly, blue-eyed, one-armed Texan with an appetite for violence. A sometime logger, mechanic, gunsmith and draughtsman, he was amongst the most dangerous and feared men in the state. Despite his disability, he was a ferocious fighter and had proved himself more than capable of taking on six men simultaneously in a brawl.

But Higgins, like many another of his kind, got religion. And there was anyway much, much more to this driven man than a taste for bare-knuckle bruising. Always more than the sum of his – somewhat diminished – parts, Higgins loved books and reading, and was a dreamer. When he reformed, the brawler became a Baptist Sunday school teacher in his Beaumont, south-east Texas, home.

And he also became obsessed with oil and the notion that it could be found in large quantities in an area where he had discovered seepages. For Higgins had been leading a Sunday school outing to a hill called Spindletop when he found some small springs through which gas was bubbling. When he poked a stick in the ground, he was able to light the gas escaping from the hole.

The next move was to form an exploration and production company, and in 1892 the Gladys City Oil, Gas & Manufacturing Co. (named, naturally, after one of the girls in his Sunday school class) went to work.

It found promising signs but nothing else. Even though money was a constant and worsening problem, Higgins refused to give up the search. He *knew* oil was there, deep underground, waiting to be found. As the years went by, Higgins was frequently denounced as crazy; even his own doctor told him that he was a victim of hallucinations. Various geologists Higgins brought in examined the area and left having added their voices to the chorus of jeers.

In a final act of desperation, Higgins – who had by now been forced to sell almost everything he owned to finance the quest – advertised for a driller.

Captain Anthony Lucas, born in Dalmatia of Slav parents, was a graduate mining engineer who had studied in Graz before emigrating to America. He was also the man who developed the theory that oil deposits could be found at the sides of subterranean salt domes.

In mid-1899, Lucas signed a lease and option agreement for the mining rights on the Gladys City land and another 33-acre lot Higgins owed money on and had thus been unable to sell. Under the terms of the deal, Higgins had a 10 per cent interest.

Lucas began drilling immediately and continued until January 1900. Although he failed to strike, he had seen signs of oil. But now he too had exhausted his capital. Playing his last card, Higgins (assisted by Lucas's wife) persuaded the engineer to try and interest somebody else. Lucas went to Guffey and Galey, the best-known wildcatters in the business, operating out of Pittsburgh.

Guffey was the businessman and drove hard bargains. He and Galey would take on the work but would also take seven-eighths of the proceeds in exchange for their backing. Lucas would have one-eighth of the deal. As for Higgins, well, if Lucas wanted to cut him in, that was his business. There was a choice: take it or leave it.

Lucas took it and Galey visited the site, looked around and said drilling should begin exactly where Higgins had first indicated, by the streams. He also said the result would be 'the biggest oil well this side of Baku'.

Work began in autumn 1900 and broke off for Christmas. Oil had shown at around 880 feet. Work resumed on New Year's Day 1901 and on 10 January mud began to bubble furiously in the well. Within seconds, six tons of drill piping had been blasted out of the hole, up and out of the derrick taking the top clean off.

There was a brief silence during which the drill crew tried to clear the debris. Then came a sustained, deafening roar as first gas, then rocks and finally oil came hurtling out of the well. The noise could be heard in Beaumont. The well, named Lucas 1 Spindletop, was flowing not at the anticipated rate of 50 bpd but at a staggering 75,000 bpd.

Patillo Higgins' obsession had resulted in the discovery of the mother of all gushers.

What followed was a kind of honky-tonk madness – the excesses of Oil Creek and Pithole merged and multiplied by 100 and then by 1,000. Spindletop was where the gaudy circus came to rest, put down roots and flourish as never before.

Here, legends were made by the day, by the hour even. Here, Mrs Sullivan's pig pasture, a plot a man might spit across with a following wind, changed hands for $35,000. Here, a single acre rumoured to hold the key to oil was sold for $900,000. Here, whorehouses mushroomed to entertain the 16,000 of the hopeful horde – whose total was being added to at the rate of six trainloads per day – living in tents on the slopes of Beaumont. Here, in this classic Texan one-horse town, the population rocketed from 9,000 to 50,000 in a matter of weeks.

And here, oil was flowing so freely and in such abundance from the new wells that sober calculations produced preposterous results: could Spindletop alone really produce as much as the whole of Pennsylvania? Could it really account for half the nation's entire oil output? The answer was a triumphant affirmative in both instances.

The consequences were momentous. Profound changes in industry and transport were set in motion immediately. Although the new Texan oil was heavy and sulphurous, unsuitable for refining into kerosene, it made excellent – and blissfully cheap – fuel oil for railway locomotives, ships and industrial furnaces. A railroad that in 1901 had one oil-burning locomotive was, in the wake of Spindletop, running 227 within four years.

Such was the extent of the glut that by mid-1901, oil was selling for three cents a barrel. Water, by contrast, retailed at five cents per tin cup in the oil town.

The first Marcus knew about Guffey and Galey's strike was when he read about it in his morning newspaper. In the US, the story had been making headlines day after day. In Britain, the event – if it was recorded at all – rated little more than a paragraph.

Neither Marcus nor Sam nor even Fred Lane had heard of Spindletop or, for that matter, Beaumont. Office atlases were of little help, wall maps even less so. Moreover, none of them had heard of Colonel James Guffey – who only rarely left Pittsburgh, it seemed – or John Galey. Contact was made only with great difficulty after many telegrams and a blizzard of letters had been dispatched. Marcus became 'intolerably impatient' during the delay. His *amour-propre* suffered serious damage when, contacted at last, Guffey and Galey said they had never heard of Sir Marcus Samuel or his Shell company. Umbrage was taken in Leadenhall Street.

By and by, after Marcus had opened negotiations and after the exchange of many – sometimes agitated – cables, a huge deal was concluded. Shell would for 21 years carry half of Guffey's production at a fixed rate of $1.75 per ton plus 50 per cent of the profits on sale with an annual minimum delivery of 100,000 tons (14.7 million barrels).

This provided freights and, of course, oil (mainly in the form of liquid fuel which delighted Marcus given his ambitions of oil-fuelling the Royal Navy) which could legitimately be sold in Europe. Marcus therefore bought a £90,000 stake in Gehlig-Wachenheim, a German company, substantially owned by Deutsche Bank, which supplied kerosene to the domestic market from oil it refined and stored in Batum. This became Shell's marketing foothold in Europe (and, Marcus believed, a crack in Standard's European defences) from which sprang a new company known as PPAG.

There were, to Marcus, two other great benefits to the Guffey deal. The first was that he believed Guffey and Galey did not get on with

Standard (they didn't, to the extent of using lawsuits to keep Standard out of Texas while, nevertheless, still selling oil to the company). Second, there was Spindletop's geographical location. Beaumont was on the Sabine Lake, only 19 miles inland of Port Arthur on the Gulf of Mexico, a fact that made onward transport a simple and straightforward business. The immense Guffey deal and Marcus's European initiative touched off a new series of intricate manoeuvres.

Standard now began courting Shell. Marcus somehow came to believe that under the terms of an alliance, Standard would permit Shell to remain an independent operator. He liked the idea sufficiently to go to New York in October 1901 for serious discussions.

At the same time, Fred Lane had restarted talks with Royal Dutch and, in November, gave Marcus the draft of an agreement between Deterding's company and Shell. Marcus, however, thought a Standard–Shell deal to be imminent and turned down Lane's proposal, countering with an idea to create a British–Dutch distribution company which would charter Shell tankers at a high rate.

Now Deterding, who had been quietly organising a combination of small producers under the Royal Dutch umbrella, thus capturing a substantial chunk of the East Indies market, found himself on the horns of a dilemma. Agreeing to Marcus's proposal would prove expensive: the proposed freight rates were significantly above the market norm. But rejection would leave his company vulnerable to being swallowed up by a Standard–Shell giant. He chose, in the dying days of 1901 and with no great enthusiasm, to accept.

Meanwhile, Standard made it clear that though they were willing to pay $40 million for Shell, with $13 million to Marcus personally, there was no way they would permit his company any kind of independent existence.

Marcus, who believed himself to have been egregiously misled by Standard, was initially furious. But he soon came to regard the affair as not much more than an irritant because, buoyed up by the great Guffey deal and Deterding's acceptance of the British–Dutch proposal, Shell's position was again looking strong. Fat and fire had once more been safely separated.

There was, too, a great civic occasion to look forward to: his election as Lord Mayor of London on 29 September 1902. And there had been a small incident of, for Marcus, enormous social significance. He had been out riding in Rotten Row when he had been approached by one of the most senior Waley-Cohens, who had a son, by the name of Robert, who had recently graduated from Cambridge with a good degree in chemistry. The young man had joined the government's meteorological service but was unhappy about his prospects in the civil service. Might Shell be able to offer something more interesting?

The Waley-Cohens were at the top of the aristocratic Anglo-Jewish heap. As little as ten years earlier they would not have given Marcus Samuel the time of day, still less sought employment for one of their bright offspring with him in Shell. It was a major milestone on the road to acceptance, and it still mattered greatly to Marcus. As it happened, Robert Waley-Cohen would become one of Shell's most able, most valued and longest-serving managers.

LOST AT SEA

The sequence of events that would bring Marcus Samuel low and take his company to the brink of collapse began with a Shell tanker, laden with kerosene, running aground in the Suez Canal. Another company tanker went to her aid and, during pumping operations aimed at lightening the first vessel, caught fire and burned out.

Marcus, at exactly that time, was seeking leave from the Canal authorities to ship petrol, refined in Balik Papan, in bulk through the waterway and on to Britain. The petrol venture had proved far more difficult technically and vastly more expensive to set up than had been envisaged. Having to send tankers round the Cape of Good Hope rather than through the Canal would add enormously to costs, but with a vessel burning within sight of their offices, the Suez authorities unsurprisingly turned down Shell's request.

The next hammer blow struck at the heart of some of Marcus's most deeply held beliefs and cherished values.

He had for years and recently with the assistance of the First Sea Lord, Admiral Sir John Fisher, been trying to persuade a deeply conservative Admiralty to switch Britain's huge fleet from coal to liquid oil fuel. Given the political clout of the coal owners and the country's huge historic investment in steam-powered transport and industry, it was always going to be a long and difficult campaign. A further and strategically vital consideration concerned supply. Britain appeared to be built on foundations of coal, but where was the indigenous oil?

Fisher, a bumptious, belligerent character reckoned to be the least popular flag officer of all time, was seeking to shake the navy out of its complacency with a series of radical reforms. He lost no opportunity to hammer home the superiority of oil. It was cheaper, much more efficient thermally, lighter – which of course yielded tactical speed advantages – and did not require gangs of stokers to feed and tend hungry furnaces all the time a warship was at sea. And Fisher made much of the ability of oil-fired vessels to operate at speed without trailing clouds of giveaway smoke in their wake.

Finally, after Marcus had offered the Admiralty seats on the board – indeed, virtual control of the company – in his attempts to assuage their anxiety over the supply issue, as well as making available all the technical information Shell had gained from operating oil-fuelled tankers, a public trial was agreed.

HMS *Hannibal* had been fitted with oil-burning equipment and, on 26 June 1902, Marcus took the train to Portsmouth together with the Admiralty's experts to watch the trial. A crowd of several hundred saw the warship leave harbour. She was, as had been agreed, at this point burning best Welsh steam coal. Marcus and Fisher, standing together, noted the wispy trail of light smoke in her wake.

At a heliographed signal from the shore, *Hannibal* switched to oil fuel – and began immediately to be wrapped in black blankets of thick, sooty smoke.

What neither Fisher nor Marcus knew was that *Hannibal* had been fitted with obsolete vapourising burners rather than the atomising kind used by Shell and other commercial operators. The vessel was now entirely lost to view as smoke in vast quantities belched from her funnels. It was an appalling spectacle, a humiliating fiasco.

In defence terms, the events in Portsmouth harbour were to have disastrous consequences. Oil-fuelling of the fleet was put back for almost ten years, contributing in large measure to Britain's lack of readiness for battle in 1914. Some historians have argued that the nation came close to losing the war as a direct result.

For Marcus, the trial was a mortal blow not only to his hopes of a patriotic partnership with the navy as fuel supplier but also of Shell being seen and valued as an agency of government, an integral part of the great imperial machine.

Disconsolate, Marcus returned to his office where he found only more bad news: a report from Australia showed that a venture he had been pursuing and in which he had invested more than £100,000 of Shell's funds had yet to yield a penny in return. Shell was again in serious and worsening trouble.

There was on his desk another document – a contract prepared by Deterding and the Rothschilds – requiring his urgent attention. On 27 June, Marcus Samuel put his name to it and effectively signed away Shell's independence and his control over it.

Earlier in the year, Fred Lane had arranged a meeting between Marcus Samuel and Henri Deterding. It was never going to be a socially easy occasion. Deterding, 13 years younger than Marcus, believed Samuel to be intolerably arrogant, strong on vision but weak, sometimes beyond belief, on detail and organisation. He was too easily – and often – distracted from business matters and was anyway far too concerned with winning the approbation of those he thought to be his social superiors.

Marcus thought Deterding clever – there was no denying that – but conceited, far too opinionated, altogether too sure of himself and obsessed with details that tended to obscure important parts of the big picture.

Lane set up the first meeting in London. Deterding, who knew of Marcus's susceptibility, launched a charm offensive. It was so effective that he was able to persuade Marcus that their agreement of 27 December, potentially so expensive for Royal Dutch, had yet to be ratified by the board in The Hague and was thus not binding.

In fact, Deterding was angling after not only a way off the penal freight rates hook but also an entirely new arrangement, the ramifications of which would be enormous. At their second meeting, Deterding proposed the formation of a completely new joint marketing company to handle activities in the Far East. Of course, a man of Sir Marcus's stature, experience and reputation could not be expected to occupy a back seat.

'But,' said Deterding in a blinding flash of self-revealing description, 'mine is a personality which does not readily submerge itself.' This would later come to be regarded as understatement of truly heroic proportions. At the time, 6 April 1902, it led to both men agreeing to make the most of their different abilities. Marcus thus became chairman and Deterding, charged with responsibility for the company's day-to-day activities, managing director.

On 23 May, Marcus signed a new draft contract with Royal Dutch. But Deterding, again almost before the ink had dried, sought drastic revisions. The cause this time was the emergence of what he regarded as an excellent, and probably unrepeatable, new opportunity in Europe. It had come about, Deterding explained, by the presence of Marcus in Europe through PPAG. Given this, he wanted to draw into the British–Dutch alliance none other than the Rothschilds.

Deterding added that he wanted all three parties to have equal holdings and status in the new company – still under the chairmanship of Sir Marcus and, of course, his managing directorship – so that Europe could be dominated by Texan oil from Shell and Russian oil from PPAG, and the Far East sewn up with Royal Dutch and Shell oil from Sumatra and Borneo and Bnito's Russian oil.

It was a sophisticated proposal of the complexity, in the numbing detail, for which all Deterding-inspired arrangements would become notorious. And it was undeniably clever. But above all, it was sound. There was more.

Deterding had personally acquired concessions in the new Romanian oilfields. The major Romanian company was Steaua Romana, controlled by Deutsche Bank, of which, through Gehlig-Wachenheim, Marcus was already an associate. Romanian oil could, self-evidently, be sold much more cheaply throughout Central Europe than Texan or even Russian oil – something that would disoblige Standard mightily. How bad could that be?

If Deterding was expecting applause or gasps of delight and admiration from Marcus – and there was, after all, a splendid symmetry about the idea – he was sorely disappointed. Marcus attempted to elevate a clutch of minor matters to the status of reasoned objections.

What Deterding either did not know or chose to ignore was that Marcus disliked doing business with the Rothschilds and had a serious objection to doing business with Romania. For, even as they were speaking, Jews were being beaten, imprisoned and even killed in a vicious Romanian campaign of persecution. Moreover, Marcus had already planned the route of his inaugural parade as Lord Mayor through Portsoken, the Jewish quarter of the East End where he had been born and where Romanian Jews were at this moment seeking refuge.

Nothing was agreed or signed at the meeting. Deterding returned to Holland stunned and angry at the display of 'heart over mind' thinking he had witnessed. He had already suffered one nervous breakdown – caused in early 1902 by his frantic efforts to disentangle Royal Dutch from the hugely expensive chartering contract with Marcus – and Lane was anxious that he shouldn't have another. He wrote quickly to Deterding: '. . . you must have patience; on no account must you worry yourself because you will unfit yourself for business again'. In truth, Lane was himself becoming increasingly concerned about many aspects of Shell's affairs.

On 25 June, Shell shareholders gathered for their AGM (there had been an extraordinary general meeting earlier in the year at which, among other changes, Marcus had proposed the raising of another £1 million in capital with 100,000 preference shares of £10 at 5 per cent per year) and were given some new numbers. Shell was already a major multinational with 31 installations in China, Japan, India, Egypt and Australia and another 11 under construction. There were no fewer than 320 subsidiary depots in eastern hemisphere towns amounting to £3.5 million in tangible assets. The strategic plan now was to get production as near to customers as possible and create business in Europe, South America, the West Indies and South Africa on a similar scale to that in the East.

Everything was approved at what was generally a downbeat meeting, at which shareholders sensed their company was in trouble: profits had fallen sharply and the declining price of oil made quick recovery an unlikely prospect. Their mood would have been even more depressed had they known what was about to happen the next day with HMS *Hannibal* in Portsmouth.

But there was more and infinitely worse news on the way. Spindletop was about to become Swindletop: grotesque overproduction and crass mismanagement – wells drilled so close together that one area was known as 'the onion patch' – had lowered the oil-driving gas pressure in the field to such an extent that Guffey's wells were running dry. Shell's western lifeline had snapped. Four of the big tankers intended to handle Texan oil were converted into cattle carriers and used to service the east London meat trade.

At the same time, the eastern lifeline – the anticipated charters by the Asiatic Petroleum Company so recently and painfully formed – was fraying under strain.

The problem was simply stated. When Asiatic had been formed with, as Deterding had proposed, Shell, Royal Dutch and the Rothschilds as equal stakeholders, Marcus believed that it would mean the end of the bitter little sales war between Shell and Royal Dutch. Also, the writing into the contract of the terms of charter for Shell tankers would surely guarantee employment for at least part of Shell's fleet in the East just as Guffey's oil did in the West.

But here, in black and white, was another startling instance of Marcus's inattention to detail proving disastrously costly. Shell and Royal Dutch were indeed linked but *only* in the Far East. As Deterding had understood perfectly, they were competitors in every way everywhere else. And Deterding had also understood perfectly that the newly constituted Asiatic had absolutely no obligation to use Shell's tankers. Indeed, prudent management of Asiatic's affairs dictated that Deterding should at all times charter vessels as cheaply as possible. And not even Marcus could claim that Shell's charter rates were of the bargain basement kind.

Deterding had moved swiftly to get Asiatic up and running from offices at 21 Billiter Street in London, literally next door to Standard (registered in Britain as Anglo-American). Besides himself as managing director and Marcus as chairman, Asiatic named Fred Lane as deputy managing director, J.Y. Kennedy as secretary, Robert Waley-Cohen as assistant manager and S. Hodgkinson-Lowe as cashier. Nobody in the oil business was under any illusion about what was happening and who, precisely, was making the big decisions.

Marcus turned away from Shell and business now, for he was about to take office as Lord Mayor. His parade – despite objections from the police that they had too few officers to line the route effectively – passed through the Portsoken ward Jewish quarter to the appreciative cheers of those who had literally seen nothing like it before. Deterding, who was in attendance as Marcus's guest, was unimpressed: it was 'not worth a white tie to attend a second time' he wrote to a colleague, adding: '. . . to Dutch eyes it was more like the ceremonial parade of a circus'.

The Lord Mayor's heavy programme of civic duties kept Marcus away from his office for more than nine weeks. It also took a heavy toll on his health. In addition to constant, debilitating headaches, he had to have all his teeth removed.

When he was able to 'put in an hour at the office' in Leadenhall Street, Marcus found only bad news awaiting him. On consecutive days, he was told by Sam that earnings were so poor that a dividend of only 2.5 per cent could be paid, and that from making a profit of £1,000 a day in 1900 they were now losing £500 a day.

Two days after Christmas 1902, Marcus at last found time to read a letter from Fred Lane that had been lying in the office.

The letter had the explosive impact of a grenade. A friend and trusted partner for more than 20 years, Lane was quitting the board of Shell – and not because of the extra demands on his time made by his new Asiatic responsibilities. He had seen and knew too much to go quietly and, in a carefully constructed and devastating letter, laid bare the reasons for the mess that Shell had become and which now, indeed, might make mere survival seem a bold ambition:

> You are, and always have been, too much occupied to be at the head of a great business . . . There seems to be only one idea: sink capital, create a bluster, trust to providence. Such a happy-go-lucky frame of mind I have never seen in business before.

Lane found recklessness in too many instances. He had encountered crass incompetence, woeful ignorance and an extraordinary unwillingness to hire competent staff or even, at times, any staff at all. Financial control was laughable and hundreds of thousands of pounds had been squandered. 'Business like this,' Lane wrote, 'cannot be conducted by an occasional glance in one's spare time, or by some brilliant coup from time to time. It is steady, treadmill work.'

It was Lane's damning judgement that if 'treated on its merits, the company would undoubtedly have to be written off as a failure. A great splash has been made and the situation capitalised; but it cannot last, and the bubble will burst.'

Astonishingly, Marcus thought the damage could be repaired with, and Lane placated by, a letter. It proved otherwise. The two men met and talked at length. There was at least one more exchange of correspondence. Lane, insisting that he was not seeking to evade his responsibilities and that his counsel and experience would always be available to Marcus, declined to reconsider or withdraw his resignation. Each side blamed the other and even basic civilities were imperilled as the bitterness increased. Lane went. Both men believed themselves to have been betrayed.

DECLINE AND FALL

When he looked at the business scene now, Marcus was confronted by a stark contrast: Shell, palpably holed below the waterline, was limping from crisis to crisis; Asiatic, under the dynamic Deterding, was fairly flying.

Deterding had begun trading before the detailed establishment of the Asiatic Petroleum Company had been completed. That had finally been achieved with the agreement of ten contracts in May 1903. But it had not been done easily or without some incendiary rows. Precisely what happened is disputed, but at the core of the conflict was a battle for control made memorable by the clash of monumental egos. The Royal Dutch view, as articulated by official historian Dr F.C. Gerretson, is that Deterding was seeking – vigorously – that in a company composed of three equal partners, everybody should act 'rightly and fairly'.

Marcus Samuel's biographer saw things differently. Such was Deterding's determination to get his own way, he was 'driven into a state of unreasonable rage and unreasoning venom' and was, indeed, at one stage 'close to dementia'.

Given that Deterding wrote letters at this time portraying Marcus as 'a rascal with a knife' and himself as a victim of stabbings and knifings, that judgement seems sound. There are two other significant points about these documents: the handwriting – a violent, scrawled assault by pen on paper – is described as being obviously the work of a man in an apocalyptic rage. And the letters that provoked these ink-splashed, heavily underscored and much-amended replies have never been found. Gerretson believed Deterding destroyed them.

The outcome was surely never in doubt. But even Deterding's most fervent admirers must have been surprised at the scale of his victory. Almost the last battle of importance concerned the duration of his term as managing director. Marcus sought to impose a limit of three years. Deterding – supported, as so often in these skirmishes, by Lane – said: 'Twenty-one years and not a day less.' He prevailed over that, too.

Staring at defeat on all fronts now, Marcus tried to salvage something at least from the disaster in Texas. He called for legal opinions on whether he could sue Guffey for failing so spectacularly to supply the oil he had contracted for.

An American lawyer examined the original contract and, in an echo of Fred Lane's strictures, pronounced it 'incredibly neglectful'. An action in Britain might succeed, but . . .

In fact, the mighty Mellons, the US banking family who had financed to the tune of $300,000 the Spindletop wildcat and had loaned Guffey several million dollars more to get the field into production, were also sorting through the wreckage.

Clearly, the contract with Shell was a major obstacle to a complex recovery operation in what would anyway be a heavily depleted oilfield. Andrew Mellon, Guffey's chairman, went to The Mote to talk to Marcus. It was a high-risk tactic: by setting foot in Britain he laid himself open to being sued and, quite possibly, ruined. But Mellon believed Marcus to be a reasonable man and his gamble paid off. By the end of his visit, Guffey was off the hook.

Now, with his year as Lord Mayor over and Shell's position weakening almost by the minute, Marcus faced an assault by a press made keenly aware that all was far from well by belated accounts and the publication of incomplete reports. On 15 December 1903, for instance, the appallingly delayed 1902 report was published but with accounts that were 'only provisional and subject to revision when figures from Asiatic should be received'.

The *Daily Mail* told its readers:

> The pessimistic rumours about Shell are unfortunately borne out by the very unsatisfactory report published . . . After paying the dividend on 1 January, there is only a carry forward of £19,556. Altogether, it is not a report of which Sir Marcus Samuel and his co-directors need feel proud.

The Financier was no less harsh: 'However the directors may try to excuse themselves, they cannot escape the reproach of exceedingly poor management.' In a commentary that might have been written by Fred Lane, the paper added:

> . . . if the report submitted in June 1902 was bad, that now presented is a good deal worse. It seems an extraordinary thing that although the company has had twelve months in which to prepare its report, the document is so incomplete.

And it was about now that Marcus's woes were added to by another rejection from the Admiralty.

He had earlier complained to Deterding that Asiatic was, to Shell's detriment, discriminating against oil from the Kutei wells and the Balik Papan refinery in Borneo.

True, said Deterding, but for sound technical and commercial reasons. The Kutei oil contained such a high percentage of heavy hydrocarbons that even when it had been refined by expert Dutch technicians, it was still of unacceptably low quality when compared to available alternatives. Moreover, after only a limited period of storage, the refined product quickly reverted to a brown colour, which could only be removed by further, expensive processing.

It was Robert Waley-Cohen, the man with a good Cambridge chemistry degree, who persuaded Marcus to have the Kutei oil analysed by the university's chemistry department.

The analysis was carried out by Humphrey Jones, later to become a key member of the Royal Commission on Fuel and Engines, and showed the oil to be exceptionally rich in aromatic hydrocarbons. This made it unsuitable for kerosene but excellent for gasoline with a remarkably high by-product of toluol, a chemical essential in the manufacture of tri-nitro-toluene (TNT) and sought by every government and explosives manufacturer.

To a navy-loving patriot like Marcus, the next move was obvious: through Fisher, he immediately made contact with the Admiralty and offered to supply them with TNT or, if they preferred, Borneo crude from which they could make TNT in a plant of their own.

Their reply – then and astoundingly and with nearly fatal national consequences again in 1914 – was a resounding negative.

Toluol, the desk sailors said, when derived from oil was but poor-quality stuff. Toluol of the quality they needed for TNT was derived from coal tar in special, secret plants at certain gas works. It was true, they added, that only a limited amount could be produced this way, but it was perfectly adequate in peacetime and nobody, with the possible exception of the tiresome Fisher, was anticipating a war.

The German government, however, had seen things differently. Krupps had already built a plant for manufacturing TNT from Borneo crude supplied, as it happened, by Shell through PPAG.

Later, Marcus persuaded the Rothschilds to take up the matter with the French government. The result was an order from the French army and navy of a size that demanded the building of a British–Dutch plant in Rotterdam. That plant was to figure prominently in Marcus's elevation to the peerage. It was also to prove every bit as important in saving Britain's bacon at a time of great peril.

And still the problems piled up. Standard of course knew of Shell's decline and imminent fall but could not resist having another go at driving a stake through the heart of Marcus's company. They built an oil facility in Romania which, according to a *New York Herald* story, 'delivered a crushing blow to Shell . . . [Mr Rockefeller] has well nigh driven the Shell company to the wall'.

Marcus sued for libel, got the statement withdrawn, an apology and his costs. But that mattered less than the fact that the story was essentially true.

But then came the bitterest and most humiliating blow. Confronted by the certainty of Shell's losses escalating during another price war with Standard made possible by their new Romanian facility, Marcus wanted to withdraw from PPAG and the whole of Europe if only Deutsche Bank would let him.

The bank agreed and said they would take over Shell's holding in PPAG at par as long as Shell sold to PPAG six of its best tankers not at cost price but at 'book' – fully depreciated and written down – value. Marcus would have to let the vessels go at a huge loss.

There was, of course, no choice. A beaten man, he would be obliged to watch as the best of the Shell tanker fleet, his own marvellous idea translated into revolutionary reality by Fortescue Flannery, sailed into new ownership at a ridiculous price and under a German flag.

There was now only one way in which something – anything – of Shell could be saved. Marcus went to see Deterding and suggested a Shell–Royal Dutch amalgamation on the same basis as the Asiatic arrangement – 'no party to have any advantage over the other'.

Deterding refused even to consider such a deal. The numbers made no sense. Shell, dependent on Deterding's management of Asiatic, had struggled to pay a 5 per cent dividend. Royal Dutch, by virtue of Asiatic, had paid dividends of 65 per cent in 1903, 50 per cent in 1904 and 1905, and in 1906 would pay 73 per cent. With these figures in mind, the only arrangement Deterding would discuss was an amalgamation in the ratio 60:40 in favour of Royal Dutch. It would, of course, mean that control of Shell would pass into foreign hands.

Over the winter of 1905 and well into spring 1906, Marcus sought with increasing desperation to wring better terms out of Deterding and Royal Dutch. *The Times* had meanwhile weighed in with a lament at the sight of Shell tankers flying a German flag and a condemnation of the government for 'parting with a vital war interest' – an issue obvious to everyone except, it seemed, the Admiralty and the government. There was no discernible movement in The Hague.

In April 1906, Marcus went back to Deterding in Billiter Street, where he was courteously received. Deterding laid out his last and best terms. Shell would be a holding company, owning no more than 40 per cent of a new Royal Dutch/Shell group. The holding company would, however, remain under the control of Marcus, who would retain the special 5:1 voting rights on his own shares.

To ensure that Royal Dutch would run the group to Shell's benefit, it would buy a quarter of the shares at a premium. In the past three troubled years, Shell's shares had fallen in value by more than 60 per cent. Now, instead of the current quoted price of 23 shillings, Royal Dutch would pay 30 shillings for 25 per cent of the stock.

When he reported back to London, Marcus's colleagues on the Shell board still hesitated. Marcus told them he would go back to The Hague in the hope of bringing 'satisfactory news' to the next meeting. He also told them it was unlikely that Royal Dutch would improve the deal.

He was entirely right. The offer still stood but now he had to end the procrastination and make a decision. 'I am at present in a generous mood,' said Deterding. 'I have made you this offer, but if you leave this room without accepting it, the offer is off.'

Marcus accepted. But still the pain and the humiliation went on. On 1 June 1906, he had to face formal censure at an Asiatic Petroleum Company board meeting. Deterding moved a resolution 'condemning recent correspondence conducted without the approval of the executive committee and detrimental to the company's interests' between Marcus and some company agents. The motion was seconded by Fred Lane and put to the test. Five members voted in support, three – Marcus himself, his brother-in-law Henry Benjamin and newly appointed director Robert Waley-Cohen – voted against. Marcus was thus censured.

The hurt turned now into bitterness and smouldering rage. Marcus had arrived at what must have been the lowest point of his long career. At least the wretched business of the censure had been conducted in private; the finalisation of the amalgamation, and what Marcus regarded as his deeply humiliating loss of executive control to Deterding, was necessarily carried out in public.

Particularly wounding for Marcus, for whom the chairmanship of the newly allied companies was a feeble substitute for a proper decision-making, big-stick-wielding role, was the enthusiasm with which the new arrangements – including his own drastic downsizing in importance – were greeted by the financial and trade press.

Marcus made only one comment at the time about the effective loss of the company and global business he had worked so hard to create. 'I am,' he said in a simple statement to the *Daily Mail* that was much quoted on both sides of the Atlantic, 'a disappointed man.'

Deeply disgruntled, he awarded himself the *Lady Torfrida*, a magnificent 650-ton motor yacht and another token of tycoonery, and quickly set off on a long cruise. Marcus loved the sea and was always obviously and abundantly happy on his yacht. Sadly, Fanny was a poor sailor and was never able to share his enjoyment in anything except the calmest of conditions.

Back on shore, the wounds were still evidently raw. When Marcus announced – again in an interview with the *Daily Mail* – that he had decided to retire from the managing firm of M. Samuel & Co., he delivered himself of views that must have caused acute discomfort all over Whitehall and Westminster:

> I should never have retired if I could have found among contemporary statesmen any man of the calibre of Lord Beaconsfield [Benjamin Disraeli] who placed a government representative on the board of the Suez Canal Company, and who would have taken similar action in the all-important matter of retaining under British control and guidance the greatest oilfield for liquid fuel in the world – that of Borneo.

He was, of course, referring to the Admiralty's rejection of his June 1902 offer to put government appointees on the board of Shell.

Sir Marcus went on:

> But we live in degenerate days and the men at the head of affairs, however high-sounding their names, are mediocrities, never looking beyond tomorrow, afraid of responsibility, and utterly lacking any business experience.
>
> Sir John Fisher is the only man I have found with any backbone

> . . . I do not see the use of contriving work for the mere purpose of money-making, having long ago realised as much as I need. The only reason for going on would be to do a great public service.

The shareholders of Shell marked the retirement of Marcus by subscribing to a portrait of him by Sir Hubert von Herkomer. The painting was presented to him at the Savoy by Sir Fortescue Flannery, who, on a night of fulsome speeches, paid handsome tribute to the pioneering vision of Shell's founder.

Sir Marcus, however, having said he 'looked forward to enjoying the comparative leisure which lies before me', breathed fire over the comfortable company. An unapologetic high Tory capitalist with nothing but contempt for anything faintly redolent of liberalism, he lashed out at:

> The most extraordinary delusion of the labouring man in claiming a very large share of the prosperity he has done nothing to create . . . if it were not for men with the brains who were willing to risk their all on such developments [as Shell], where would the British workman be?

Now, away from the City and the affairs of a giant corporation, Marcus sought and found consolation in the country and out on the water. He went fishing with his beloved Fanny in the beautiful, tree-fringed lake set in the grounds of The Mote. Marcus had no interest in casting flies at trout from the manicured banks of an exquisite southern chalk stream or at lordly salmon in a mighty Scottish spate river. He and Fanny angled for roach and perch from a small boat. If their luck was good, there might be another pike from a lie beneath a leafy overhang where Marcus had once caught a jack of four pounds.

But the phrase 'comparative leisure' was well chosen by Marcus because he was still, of course, chairman of Shell, of Asiatic and of Anglo-Saxon Petroleum, and a director of Bataafsche in Holland and Allied Assurance in London.

Anglo-Saxon? Bataafsche? Indeed, it was at this defining moment in the Shell story that an already elaborate corporate structure was turned into a sort of *Through the Looking-Glass* filigree of fabulous complexity and dottiness.

For the first thing that must be said about the Royal Dutch/Shell

Group of Companies ('the Group') that emerged from the great amalgamation is that it was, and has long remained in one important sense, unreal. The Group has never existed as a legal entity.

Neither Royal Dutch nor Shell Transport ceased to exist when they were formally allied on 1 January 1907. Though they merged their interests, they retained their separate identities. Each thus became a holding rather than an operating company.

Income-generating activities – the finding, producing and selling of oil and gas – were, and still are, carried out by a raft of operating companies of which the Anglo-Saxon Petroleum Company in London and Bataafsche Petroleum Maatschappij (always known as 'the Bataafsche') in The Hague, were the first.

Specially created for the task, they took over almost all the assets of their holding companies, Royal Dutch and Shell Transport. Anglo-Saxon owned and ran the transport and oil-storage facilities. The Bataafsche owned and ran the oilfields and refineries. Both companies were wholly owned by the holding companies in the 60:40 ratio established by the terms of the amalgamation.

The original two-tier structure has, of course, mutated over the years to become still more like a club sandwich constructed by a deranged chef. For the two holding companies, Royal Dutch and Shell Transport, became 'parent companies', beneath which on the organisation chart were inserted *three* holding companies – Shell Petroleum Company Ltd in the UK, Shell Petroleum NV in the Netherlands and Shell Petroleum Inc. in the US.

And where there were two operating companies, there are now scores of dozens, all separate legal entities, in more than 100 countries. Some are wholly owned, others are part owned. Their structures are, of course, subject to evolutionary change; some, such as Petroleum Development Oman (PDO), have no obvious connection with Shell and are 55 per cent or more owned by the governments of the countries whose gas and oil they are exploiting. PDO and others like it are sometimes referred to as 'Shell-managed' or 'Shell-operated' companies. All of them, a multitude in anybody's language, make up the Group.

So far, so bad. But there is more. Because Shell Transport and Royal Dutch retained their separate identities, they also retained their separate headquarters in London and The Hague.

Common financial and commercial affairs were to be handled in London while technical matters were to be dealt with in The Hague.

Because the Group didn't exist, there could not, of course, be a Group board of management. There would, instead, be directors sitting in both The Hague and London (to say nothing of the hundreds sitting in operating and service company boardrooms across the world). Sitting on top of them, at the pinnacle of the heap, was a committee of managing directors.

All kinds of reasons have in the face of corporate and private shareholder demands for radical reform been advanced by Shell for the retention of such a bizarre structure. It used to be said that the organisation was consonant with the way Marcus Samuel and Henri Deterding did business – preferring at all times amalgamations in which the dignity and goodwill of those being swallowed up was retained through partnership and profit-sharing. Both men believed this to be ultimately far better for the bottom line than Standard's way of achieving growth by grab or by battering competitors into submission. 'To crush a rival is to make an enemy,' Deterding wrote in his autobiography.

But down the years, the duality has looked – particularly to rivals – increasingly like a crude device enabling Shell to claim either British or Dutch nationality according to taste or whichever seemed most likely to yield advantage. 'It is,' as an exasperated Third World oil minister observed during a break in negotiations, 'a truly two-faced company.'

'And there's another big plus to an organisation like ours,' said a senior Shell manager moodily pecking at a piano in his home. 'The buck can be kept in motion for years between the two HQs.'

The current and long-running reserves crisis has served to highlight the enormous price exacted by Shell's obdurate resistance to structural reform over the years, which, as outlined in the epilogue, was finally overcome in the autumn of 2004.

For if the company's multitudinous critics are united about anything, it is that an archaic, anarchic management organisation featuring twin boards and headquarters, and a committee of seven managing directors, has been chiefly responsible for a tardiness of response and what an analyst quoted by the *Financial Times* called 'lamentable decision-making'.

And that is a judgement both Marcus Samuel and Henri Deterding would surely have found repugnant.

CHAPTER 4

THE ODD COUPLE

After a little more than a year in 'retirement', Marcus was seen more often in his London office than the one day a week he said he would devote to business. And, by degrees, he began to develop a strangely symbiotic but effective working relationship with Henri Deterding. They made an odd but highly successful pairing at the top of Shell.

Having moved house from Portland Place to an even grander and more imposing home just off Piccadilly in Hamilton Place, Marcus slipped easily into his new role as the – relatively familiar – public face of Shell. Though primarily a figurehead, Marcus had solid asset value to Deterding through possession of qualities he both admired and could put to work.

First, there was indeed 'the good name of Samuel' that Marcus had further enhanced during his term as Lord Mayor of London. There was, too, his dyed-in-the-wool Britishness, although this, and the nature of Shell's allegiance under what was now 'foreign' ownership, were called sharply into question by Winston Churchill in a parliamentary attack on 17 June 1914. Churchill's remarks and a series of bitter exchanges with Sam Samuel, now a Tory MP for Wandsworth, were characterised as crude 'Jew-baiting' by Watson Rutherford when he spoke later in the debate.

And of course Marcus knew, and was known by, everybody of consequence in oil, banking and shipping. Keen to nurture as well as exploit these qualities, Deterding was punctilious in always referring deferentially to Marcus in public as 'our chairman' and in consulting him on major developments.

For his part, Marcus quickly realised that in Deterding, Shell had in the driving seat a supremely gifted executive, a manager of 'genius'. Although some of Deterding's boardroom colleagues might have winced at the description, it was one Marcus used, without irony or reservation, many times.

There can be no doubt that Deterding was rapidly establishing himself as the shareholders' darling. The new and vastly higher levels of performance and profitability being achieved under his regime were quickly reflected in the balance sheet and dividends, and the company was growing at an electrifying pace.

By now relishing his elder statesman role – to say nothing of the multiplying wealth that enabled him to buy and put into trust for his family almost 20 acres of Mayfair round Berkeley Square – Marcus had the satisfaction of seeing many of his predictions realised.

Motoring and motorcycling were growing at an extraordinary pace in industrialised countries, driving up demand for petrol at exactly the right time. For sales of kerosene illuminating oil, on which so much of the industry had been based, were falling rapidly as electric lighting became available at the flick of a switch. And soon, petrol-powered aircraft would be taking to the air in commercially significant numbers.

At sea, even 'King Coal's' most ardent admirers in the Admiralty had been obliged to admit that Marcus and his 'oil maniac' friend Fisher had been right: oil-fuelled warships were better in every respect. But having narrowly avoided defeat in the 1914–18 war, the Admiralty still delayed until 1921 an announcement that they would order no more coal-fired vessels.

And it had taken another great – and nearly terminal – crisis to focus the Admiralty's attention on the matter of Borneo crude, toluol and TNT that Marcus had been trying to interest them in for almost a decade.

Despite the Admiralty's assurances that their 'secret' coal-tar plants at gas works could produce enough toluol for the nation's TNT needs, at the end of 1914 – and after only four months' fighting – Britain's armed forces had almost run out of high explosive. It would be hard, if not actually impossible, to imagine a more demoralising situation for men facing an abundantly supplied enemy on land and at sea.

Shell produced a daring solution to what was a critical problem. On the night of 30 January 1915, each part of the company's toluol refinery in Rotterdam was numbered and the entire plant disassembled and packed into crates. The following night, the freighter SS *Laertes* crept out of port and carried the crated refinery to Britain, where it was unloaded in London and, on a specially cleared rail network, taken 150 miles by train to Portishead in Somerset. There, the numbered pieces were put together like a giant jigsaw and, within a matter of weeks, toluol produced by the plant was passed to a new Shell nitrating factory at nearby Oldbury-on-Severn, where 450 tons of finished TNT was manufactured monthly.

Later, another pair of similar factories was built and, by the end of the war, 30,000 tons of TNT – 80 per cent of their needs – had been supplied to British and French forces.

It was a brilliant idea, brilliantly executed. But precisely whose brilliance was responsible is disputed. Marcus Samuel and Mark Abrahams both professed authorship of the plan. Robert Waley-Cohen clearly played a vital part and Henri Deterding was claimed by some to have devised and masterminded the scheme. For sure, Deterding was of crucial importance in dealing with the (officially neutral) Dutch government and the Bataafsche, without whose full but clandestine cooperation the venture would have collapsed.

Ultimately, 'the Allies floated to victory on a wave of oil,' according to Lord Curzon, a member of the War Cabinet and Foreign Secretary in 1919. But maintaining supplies in the face of mounting tanker losses due to German U-boat and surface attacks had become a desperate business by the summer of 1917.

Diplomatic cables to Washington in July said that the Royal Navy would be immobilised, the 'fleet out of action', unless more tonnage was made available urgently. By the autumn, France was in an even worse position: Fuel Minister Henri Bérenger warned Georges Clemenceau, the 76-year-old radical Premier whose government was the first to use the phrase 'total war', that the country would run out of oil by March 1918.

The situation was relieved only by the formation in February 1918 of the Inter-Allied Petroleum Conference to pool, coordinate and control all oil supplies and tanker shipping. The organisation proved highly effective at distributing supplies, but it was Shell and Standard, the big two of the international oil business, who made the system work.

Shell was anyway the biggest supplier to the British Expeditionary Force and was, indeed, until mid-1917 the sole provider of aviation spirit to the Royal Flying Corps. Of the company's contribution to the great struggle, Henri Bérenger said simply: 'Without Shell, the war could not have been won by the Allies.'

It was in recognition of this, and the outstanding service rendered by each man personally during the war, that in 1921 Marcus Samuel was ennobled as Lord Bearsted, while Henri Deterding and Robert Waley-Cohen received knighthoods.

For Marcus, who shortly before his death would be advanced to a viscountcy, elevation to the peerage represented final and splendid proof that he had, at last, won the acceptance he craved. But the long journey from a Jewish boyhood in the East End to a place among the ranks of the highest in the land had taken its toll. Now 67, his health was failing and his illnesses were becoming both more frequent and severe.

It was at the AGM of 6 July 1920 that, for the last time, he reviewed the previous year's performance and also the barely believable growth of Shell from the modest but ambitious business he had formally launched in 1897 into a mighty multinational generating vast income and profits.

And it was at this meeting that Marcus dropped his bombshell: the time had come, he said, to step down and retire from the chairmanship. *The Times* reported that there were cries of 'no, no!' from shareholders but Marcus was not to be deflected. 'The weight of this gigantic business must be carried by younger shoulders,' he said, commending his son Walter Samuel as his successor.

Many warm tributes were paid to Marcus that day, including one from his old friend Fortescue Flannery, who had remained constant through the best and worst of times. There were, the newspapers reported, tears in Marcus's eyes when he said in response to Flannery: 'I should be superhuman were I not touched by the kindly things which have been said of me, and the reception you have given me.'

In full retirement, he was soon struggling with illness. He recovered sufficiently to receive honorary degrees from Sheffield and Cambridge universities, to make short cruises in the *Lady Torfrida* and even, from time to time, to engage in debate about petroleum matters by way of letters to the press.

But he became increasingly dependent on a wheelchair for mobility when he was at The Mote, and soon after Christmas 1926 it was clear to everybody that he was now mortally ill. Almost at the same time, Fanny also fell sick and, like her husband, was confined to bed, attended by nurses, in their Hamilton Place home. On 16 January 1927, Fanny died of a stroke. Marcus never knew for by then he was in a coma. He died within 24 hours of his wife.

On the morning of 20 January 1927, a quiet, respectful crowd gathered in Hamilton Place. The thick fog engulfing London muffled the voices and footfalls of Jews from the poorest areas of the East End who had gathered to mourn the passing of Lord and Lady Bearsted. A double grave had been opened in the Jewish Cemetery, Willesden, next to the tomb of a 31-year-old company commander of the Royal Kent Regiment who had been killed leading his men into action in June 1916. Marcus and Fanny were buried next to their younger son Gerald.

THE FLYING DUTCHMAN

Marcus Samuel had called Henri Deterding Shell's 'genial genius'. After his first full year in control, few – and least of all shareholders who had become accustomed to hanging on grimly in the hope that the plunging value of their investments could somehow be arrested – would have argued.

The numbers had already begun to tell a remarkable story. In 1906, Shell Transport's liabilities of more than £1 million had exceeded its assets by 2:1. Twelve months later, Deterding had turned the situation – and the company – around. In 1907, *all* liabilities were discharged. For the first time in its history, the company had no outstanding debts of any kind.

And there had been more, considerably more, good news for chairman Marcus to tell delighted shareholders: earnings for dividend were £1.5 million, something 'not equalled by the Bank of England'. Deterding, he added, had turned Shell 'into a great company'.

Having positioned Shell on the runway, Deterding took shareholders on the flight of their lives. Until the outbreak of the 1939–45 war, Shell's policy was to declare shareholder dividend as a percentage of the share's par value. At rock bottom, when it seemed that Shell was doomed in the wake of what amounted to the fire sale of the pride of the tanker fleet, investors had received a pitiful 1 per cent. Now, with Deterding in charge of the flight deck, dividends soared to dizzying heights: 1907, 15 per cent; 1908, 20 per cent; 1909, 22.5 per cent; 1910, 22.5 per cent;

1911, 20 per cent; 1912, 30 per cent; 1913, 35 per cent; 1914, 35 per cent.

As dividends reached undreamed-of levels, the value of shares rocketed. With par at £1 (20 shillings), shares at the 1906 low point were trading at 23 shillings. By 1910, when the directors called for a new issue, they decided that the price 'should not be less than 95 shillings per share'.

If genius really is little more than 'an infinite capacity for taking pains', Deterding unquestionably earned the title on that score alone. He was neither a conjuror nor an alchemist and he had no magic wand to wave over the accounts. Instead, the obsessive student of balance sheets, the man who delighted in calling himself 'a higher simpleton' because of his belief that only when a problem had been reduced to its simplest form could the best solution be found, focused on fine detail.

There was, for instance, the double taxation issue, a technical matter of eye-watering intricacy – and not at all to the taste of Marcus Samuel – but meat and drink to Deterding, who, quite legitimately, saved Shell £20,000 at a stroke.

As the man with the self-proclaimed 'unravelling habit' brought order out of the near chaos of Shell's tortured position, growth became explosive. In the five immediate pre-Deterding years 1902–06, Shell Transport and Trading's income rose from £219,567 to £428,146, while its assets were shown to have decreased slightly from £4.3 to £4.2 million. In the first five years of the Deterding regime, income rose from £556,002 in 1907 to £642,094 in 1911. In the first year of peace after the 1914–18 war, income rocketed to £4.67 million. The asset figures were even more extraordinary, rising from £4.68 million in 1907 to £6.1 million in 1911 and on to a whopping £25.9 million in 1919. And the next 12 months must have made shareholders even happier, for the 1920 balance sheet showed income to have exceeded £7.6 million and assets to have risen by almost £10 million to £35.3 million.

It was now that Shell became a truly worldwide company. Mark Abrahams had begun American operations in Tulsa, Oklahoma, in 1912, and a year later Shell moved into South America in what Deterding referred to as 'the most speculative venture of my life'.

Ralph Arnold, an American geologist, had identified 'a great plunging anticline, a type of structure that has yielded some of the best oilfields in the world'. And in Venezuela, Arnold was proved triumphantly right –

as, indeed, was Deterding in his decision to back Arnold's geological analysis. Deterding had great faith in geologists and almost none in chemists; Marcus Samuel put his faith in chemists and treated geologists as little more than witch doctors with attitude. That both men had sufficient corporate clout to ensure that their prejudices were backed with serious investment capital worked to Shell's huge benefit.

It would be difficult to overestimate the longer-term importance of Venezuela to Shell and, indeed, to Europe for it was with mostly Venezuelan oil that the Allies fought the Second World War on the Continent. Venezuela was also to become the world's largest oil exporter and, later, one of the key players in determining and defining the relationship between national governments and oil companies, and between producers and consumers.

Following an enforced two-year absence during serious civil strife, Shell returned to Mexico and, having formed a new company called La Corona, resumed production. With oil also coming in pleasing quantities from the Hurghada development in Egypt and, of course, from much further east too, the company was prospering mightily under Deterding.

And soon, Californian Shell, having spent $3 million in exploration over five years without a cupful of oil to show for it, would strike at Alamitos, and Signal Hill would become another Spindletop, only this time with staying power. Within a year of the strike, Shell's share of what would become the world's most productive oilfield was 6,000 bpd of top quality (easily refined into gasoline) crude. At the peak of activity, Shell would have no fewer than 270 wells on Signal Hill. But by then, the man in charge of Shell, a company that in terms of budget and power was already the equivalent of a minor European state, had begun to change beyond recognition.

When Henri Deterding, the short, dapper, ruddy-faced man with wide-awake eyes, had first moved to Britain to take up the managing directorship of Asiatic in 1902, he was invited by Marcus Samuel to The Mote. Deterding was instantly smitten with the relaxed, long-weekend, country-house lifestyle of the land-owning and moneyed classes. As a married man with two sons (swiftly dispatched to English public schools) and two daughters, he had naturally installed himself in a comfortable London home. But after a weekend at The Mote, Deterding set out to acquire what he believed to be the essential accoutrements of Englishness: tweed clothes, horses, a perfectly matched pair of shotguns and, of course, an impressive country house set in ample grounds.

He found exactly what he was looking for in Holt in Norfolk. Fred Lane, with whom Deterding had formed a sort of mutual admiration society (each routinely referring to the other as 'the cleverest man I know' and 'a brilliant businessman'), also had a country house in Norfolk and was on shooting-party terms with the squirearchy.

Deterding took to country life with the fervour of a convert. In fact, he developed such a fondness for field sports that the recently retired Marcus became increasingly twitchy and irritable about Deterding's 'gone shooting' absences from the office.

And when in 1921 Deterding was made Sir Henri by King George V at a Buckingham Palace investiture, the transformation of the strange, numbers-obsessed sceptical Dutchman who had mocked the fancy dress gallimaufry of Marcus's Lord Mayoral parade seemed complete. Here, surely, was a naturalised, newly minted Englishman of impeccable integrity and unimpeachable patriotism.

The truth, however, was starkly different, for a corrosive poison had already entered Deterding's soul. As he knelt before the monarch, he had already begun a journey during which extraordinary hubris would turn into barking megalomania and, having visited the wilder shores of European fascism, he would, in an act of breathtaking betrayal, become a hardline Nazi revered and ultimately mourned by Hitler.

The descent into madness began in Russia. The detail, as always with Deterding, is complex but the essence is straightforward: he made a huge – and, it must be said, rare – error of judgement by buying into the Russian oil industry at precisely the moment when most of the other major players were already looking for the exit.

RUSSIAN ROULETTE

Shell had, of course, been closely involved with Russian oil from the company's earliest days. But in 1912, Deterding bought the whole of the Rothschilds' Russian organisation. Because the deal was paid for with shares, the Paris-based bankers became by far the biggest shareholders in both Royal Dutch and Shell. And the deal also made Shell by far the biggest foreign player in Russia, with a fifth of the country's entire oil production.

But Russia in general and the Caucasus in particular were again in a volatile state. There had been scenes of horrific violence – in which people were burned alive – in two distinct but related outbreaks of rioting and ethnic turmoil in and around Baku, and at the height of the unrest at least half a dozen groups were engaged in street fighting of singular ferocity.

And there was also revolution – orchestrated by Vladimir Ilyich Lenin and Joseph Stalin – in the air. Given the antics of Nicholas II, a Tsar of exceptional stupidity, it could hardly have been otherwise. Indeed, Lenin and Stalin would both later describe the Caucasian uproar as 'the great rehearsal' for the main event – the forthcoming Bolshevik Revolution.

Deterding did not rush into the deal. In fact, he dithered for more than a year after the Rothschilds had first put their proposals on the table at Shell. Lane – acting yet again as the trusted go-between – was telling the Rothschilds at this point that Deterding could not be hurried. He likened him to an owl sitting, blinking, in a tree pondering whether or not to swoop.

And Deterding had much to ponder. For almost everybody in Europe knew that the oil region was the most seriously maladministered area of the grossly misgoverned Russian empire. Tsar Nicholas's own advisers warned that they were teetering on the edge of disaster because of the abominable living and working conditions in and around Baku and the tinderbox tensions between the races and ethnic groups. And there was also the matter of a world war raging just over the border . . .

For Deterding, who besides the appellation 'higher simpleton' now liked to hear himself described in one word as an 'oilman', there were further, technical complications. The original Baku oilfields were in decline. Production was falling and the technology that had served so well was now ageing and rapidly becoming obsolete. New and infinitely more attractive prospects were opening up at Maikop to the east and Grozny to the north-west.

After much deliberation, Deterding finally picked up the Rothschilds' deal and put Shell in the driving seat of the Russian industry – and also, as it turned out, in the front row of the stalls for the overthrow of the Tsar, the Bolshevik Revolution, the nationalisation of oil and the expropriation of the company's assets and interests.

Deterding was initially stunned and then apoplectic. Letters he wrote later to Calouste Gulbenkian – a fabulously rich, multilingual, independent Armenian oilman always in the thick of action and always looking for and prepared to make a deal – probably the best friend Deterding ever had in the oil business, make it clear that he believed any serious trouble from the Russian revolutionary left would be dealt with swiftly and severely by the military. And if by the remotest chance the Bolsheviks should succeed in taking over, they would fail and fall very quickly. As an exercise in political soothsaying, it was less than entirely successful.

Moreover, the Russian troubles had come on top of another, nearly as serious, blow from Romania. For in August 1916, Romania had entered the war in the Allied cause. At the time, the country's Ploesti oilfields were the only significant source of oil in Europe west of the Black Sea and production was running at almost two million tons a year.

In a campaign aimed at securing Romanian oil and corn as swiftly as possible, German forces stormed across the country killing, wounding and capturing more than 310,000 men in short order. For the Allies, the only option was to destroy the oilfields, and so derricks, pumps, pipelines, tank farms and installations of every kind were blown up and torched. Shell, of course, consented to the wrecking of its facilities and, at a stroke, lost 17 per cent of its worldwide production.

But what had happened in Romania was a largely unpredictable hazard of war. In Russia, Deterding had gambled and lost, proving the accuracy of Armand Hammer's dictum that 'when you lose in the oil industry, you lose plenty'. Now the man who had suffered two breakdowns while under stress in the Far East and come 'close to dementia' when apparently thwarted in negotiations over Asiatic with Marcus Samuel became seriously deranged.

For Deterding the 'genial genius' had been turned by the Bolsheviks' success into a snorting, occasionally raving, ultra-authoritarian anti-communist. The violence of his views rapidly became a cause for concern, not least when the Soviets, post revolution, sought to re-enter the oil export market. There was, as might be expected, intense manoeuvring on both sides of the Atlantic, during which Deterding at least twice sought – unavailingly – to exert pressure on British Prime Minister David Lloyd George not to allow imports from the Soviet Union.

It was in the course of some particularly delicate diplomatic dancing in America that an angry British official reported by cable that Deterding had 'quite lost his head' and, with commendable candour and in the plainest of language, added that Shell's chief had been characteristically stupid, tactless and at times seemed to be raging at his own futility.

Yet far from being cast down by these judgements, Deterding – seemingly fully recovered from the expropriation shock – went into 'Lord of the Universe' mode and began to strut his stuff with even greater flamboyance.

And it was, to be sure, an impressively large dunghill from which to crow. For as chairman Walter Samuel told shareholders in 1924, Shell was:

> working on a vast and worldwide scale . . . we are producers, refiners and distributors of oil; we make candles, road materials, lubricants, medicinal oils and a host of other by-products. We are even our own bankers, for we do not borrow a penny. All our enormous stocks of oil, tinplate, boring materials, etc. are bought and paid for out of our own funds. We are amongst the largest ship-owners in the world with a fleet of over 1.3 million tons.

As Deterding basked in the reflected glory, he turned up the volume of his anti-communist crusade to ear-bleeding levels. At the height of the row over Russia's return to the export market he even lectured, by cable, John D. Rockefeller – surely nobody's idea of a woolly fellow traveller – on the evils of doing business with the 'murderous anti-Christ Soviet regime'.

And Deterding's already rampant megalomania was pumped up still further when, after a decade of tortured negotiations, the Red Line Agreement which he had helped shape and define was signed at the end of July 1928.

Under the terms of this audacious deal, the major oil companies drew a red line on a map round what had been the old Ottoman empire at the time of its collapse at the end of the First World War and, organised into four syndicates, agreed to share equally 95 per cent of all the oil found within it. The remaining 5 per cent would go to Calouste Gulbenkian – the celebrated 'Mr Five Per Cent' – chief architect of the deal.

The agreement mattered because most experts believed that the world's last untapped reserves of oil would be found within the red line. Given that the area included Saudi Arabia, much of the Gulf and Iraq, the experts were largely right.

Dramatic proof of their judgement came in late October 1927 when oil was struck at Baba Gurgur, near Kirkuk, in Iraq. Technological developments had by the late '20s made gushers little more than a historical curiosity, but Baba Gurgur was different: the well produced a fountain of oil visible 12 miles away which defied all attempts at control for eight days. It was flowing at the rate of 95,000 bpd.

Deterding and Gulbenkian shared more than an interest in an Iraqi oil well. For theirs was also a symbiotic relationship which lasted more than 20 years. Gulbenkian brought to Deterding access to Middle Eastern contacts and overseas sources of finance; Deterding opened up to Gulbenkian the possibility of big-company backing and resources for some of his deals.

And Deterding and Gulbenkian also shared an interest in Lydia Pavlova Koudoyaroff.

For here was another area of Deterding's life in which there was profound change. Now seriously rich and, according to Gulbenkian, prone to bouts of 'overbearing grandeur', Deterding added to his country house and farm in Norfolk a fashionable apartment in Park Lane, a magnificent Home Counties country house in Buckhurst Park, near Ascot, a villa in St Moritz and Lydia Pavlova.

A vociferous White Russian, Lydia Pavlova had formerly been married to a general, was the daughter of Tsarist General Paul Koudoyaroff and had been pursued by both Gulbenkian and Deterding. Indeed, it is believed that the friendship between the two men came to grief over a woman described as 'strident, forceful and noisy'.

THE CASTLE CAPER

The second Lady Deterding was much in evidence when, in August 1928, her husband convened and hosted a conference of major oil company chiefs aimed at nothing less than the creation of a global price-fixing cartel in yet another bid to bring stability to a market in uproar.

Operating with the stealth of a B-52 raid, Deterding hired Achnacarry Castle, the hereditary seat of Clan Cameron ravaged by the Duke of Cumberland's army after the 1745 Jacobite rising, for the occasion. He sought, with apparent seriousness, to conceal the purpose and nature of the event by describing it as 'a fishing and shooting holiday for a few friends'. Given that his guests were Walter Teagle, known universally as the boss of Standard, William Mellon, chief of Gulf, and Sir John Cadman, head of Anglo-Persian, and that each was accompanied by a large secretarial staff, nobody was fooled.

In truth, the major effect of the Achnacarry conference was to increase public suspicion about the oil majors rather than establish price controls of any consequence. There were, to be sure, agreements on paper at the end of the 'holiday', but they were of little more than symbolic value.

Much more to the point was the reason for the conference: a crisis, triggered by Sir Henri Deterding of Shell, that was biting lumps out of the companies' profits. And if any further proof were needed of the extent to which anti-communism had come to dominate Deterding's life, it could be found here in abundance.

For when Sir Henri's bizarre blend of entreaties, pleas, threats, promises and warnings failed to deter American oil companies from buying cheap oil from 'the murderous anti-Christ Soviets', he launched the price-war weapon that he and Marcus Samuel had for so long – and so publicly – despised and denounced.

Now, with a world slump and generally depressed market conditions, Deterding's dotty war was causing dissent among his colleagues and heavy casualties in Shell's own ranks. It was Walter Samuel's task to announce 'with great regret' that staff would have to be let go in numbers in 'these difficult times'.

And Deterding's rabid anti-communism was, above and beyond the Russian fiasco and the price war he had unleashed, about to cost Shell a further fortune as well as adding significantly to the 'difficult times' about which Walter Samuel was wringing his hands.

For in Mexico in 1934, a new President, General Lazaro Cardenas, was sworn in. Cardenas, the former left-wing War Minister, took office at a crucial and highly charged time in Mexico's affairs. Oil production had been declining for several years and, with it, government revenues. An increasingly hard-pressed administration, swept along by ever more vocal calls for the nationalisation of the industry, increased taxes on oil sharply as a wave of strikes and noisy protests heightened tension.

Through its Mexican Eagle company, Shell was producing 65 per cent of the country's oil. Because of this, and the fact that the protests were aimed at overseas-owned businesses in general and American-owned companies in particular, Shell was in a strong position from which to conduct sensible negotiations with Cardenas.

But Deterding, by now a committed Nazi, saw Mexican nationalism as nothing more or less than a communist plot. In fact, like Adolf Hitler, the object of his veneration in Berlin, Deterding saw Bolshevik plots everywhere and, in Mexico, beneath every Shell oilfield derrick. Deterding's response to Mexican workers' pleas for increased wages and shorter hours, and to the government's demands for the replacement of foreign technical staff by trained locals and much increased Mexican participation in oil company management and administration, was an act of belligerent folly. Not remotely interested for ideological reasons in coming to terms with, still less trying to understand, the new government, Deterding now aligned Shell with the hardest of hard-nosed American companies operating in the country – the 'no surrender' faction – thus ensuring the hostility of President Cardenas and his ministerial colleagues.

With Deterding raving about the Red Menace at the merest mention of Mexico, Shell – as in Russia – was reduced to the role of disgruntled spectator as an all too familiar script was played out. In fact, the hapless Walter Samuel had barely finished denying that Shell had with other oil companies funded an unsuccessful attempt at overthrowing Cardenas when, on 18 March 1938, the President announced the nationalisation of the industry and the expropriation of all Shell's assets.

Some Shell people have asserted that Deterding's violent anti-communism and megalomania were exacerbated by Lydia Pavlova, but the truth is that by the time of their marriage, her husband's hatred of Marxists in general and the Soviets in particular was structural. Similarly, Deterding's Nazism is described in Shell publications such as Stephen Howarth's hefty *A Century in Oil*, produced to mark the company's 1997 centenary, as a sad but brief aberration. Yes, it embarrassed Shell and Sir Henri's old chums and colleagues, and yes, it provided the Nazis with high-profile propaganda. But it was only a short, albeit tragic, episode at the end of a brilliant career when, frankly, there were suspicions at board level that the old boy had actually gone mad . . .

That, however, is a long way from the truth and anyway raises many questions about corporate responsibility. The unpalatable fact is that Deterding arrived in Hitler's grateful arms only after a long affair with fascist dictators stretching back to the appalling Juan Vicente Gómez in Venezuela.

Here was an unspeakably cruel tyrant who, as president from 1909 until 1935, inflicted a reign of terror on his country. For Gómez, who refused to permit the establishment of a decent education system, saying that 'an uneducated, unquestioning people are a happy people', became the wealthiest man in South America thanks largely to oil and Shell.

Beyond commissioning overblown palaces with 200 and more rooms, most of which he left locked and unvisited, and assembling a collection of more than 100 limousines, Gómez liked to take his pleasure with young – very young – girls. At his death, Gómez had fathered 100 officially recognised bastards, while his torturers and death squads had mutilated and killed thousands for offences ranging from not cheering loudly enough during presidential processions to 'opposition' of a usually unspecified kind.

Officially, Shell said that the Gómez regime 'created a severe moral quandary' for the company. As a 'partial solution', Shell 'chose' to build an urgently needed refinery not in Venezuela itself but on the Dutch-owned island of Curaçao 50 miles away. According to American scholars, however, Gómez refused to have large, modern refineries built in Venezuela and made no secret of his reasons. An installation of the kind Shell wanted would have required highly skilled staff – either more foreigners who, like those already in the country, would have to be kept in compounds away from the locals or, worse and actually unthinkable, he would have to educate a cadre of Venezuelans.

And Deterding, as boss of Shell, was not in any kind of quandary about Gómez or Venezuela. On the contrary, he said:

> Government under General Gómez appeared sound and constructive . . . he has consistently insisted on fair play to foreign capital . . . through his policy, Venezuela has acquired a prestige and financial strength which the world depression has left unimpaired.

The death of Gómez, just before Christmas 1935, brought Deterding hurrying to Venezuela to ensure that Shell's interests were neither torched nor seized by a population suddenly freed from fear. And it also brought about an episode of black comedy that was revealing of the nature and depth of Deterding's paranoia. For absurdly, given that he was well known almost everywhere outside of the UK, Deterding chose to travel under the pseudonym Muller and suffered any number of crazy and embarrassing encounters with newspapermen.

The entire business of 'Mr Muller' careering about in Caracas must have sent a shiver down the corporate spine of Shell because the company, with good reason, was by now spending much time and money in keeping Deterding out of the newspapers and magazines. Their chosen instrument was a PR man named Adrian Corbett, who must have been one of the very few in his profession at that time paid to ensure that his clients received as little press as possible.

Mostly, Corbett had few difficulties with newspapers in Britain: the '30s were not the best of journalistic times in Fleet Street. At least one broadsheet editor said in a letter to Deterding's biographer that he had spiked a series of non-adulatory features on the oil business because Shell had threatened to withdraw a financially significant amount of advertising. But in Europe, Deterding's fascist sympathies were a staple of the newspapers. He was given a particularly hard time by a French woman journalist who ran a financial weekly and mixed just enough fact with her fiction (false 'scoops' were a speciality) to cause major problems – including a brief run on Shell shares – when she claimed Deterding had either died or absconded. And in Holland, the writing was on the wall – quite literally, for 'Death to Deterding' graffiti adorned brickwork in many parts of Amsterdam.

For Deterding, caught off guard or simply offering the world at large the benefit of his views, had become a spectacularly loose cannon prone to saying 'democracy is the lazy man's Elysium', adding that 95 per cent of the working population were mostly concerned with avoiding serious toil. And there was nothing Corbett could do to prevent Deterding from publishing his charmless memoirs, a testament to megalomania and PR catastrophe chiefly memorable for the phrase 'I WOULD SHOOT ALL IDLERS' expressed in exactly that way.

FLOWERS FROM THE FÜHRER

Of much greater concern to the British authorities was Deterding's backing of Franco in Spain, Mussolini in Italy and, of course, Hitler in

Germany. The measure of that concern is the fact that he was kept under surveillance and had his Mayfair telephone tapped for years before he was finally obliged by Shell to step down. The nature and degree of the support Deterding gave to those regimes is the subject of several internal security service (MI5) files. That the bulk of the material is still deemed – after 70 years – too sensitive to release even in summary form tells its own story.

Deterding made no secret of his idolisation of Mussolini; he was slightly more circumspect in his veneration of Hitler. In 1934 and at the age of 68, Deterding met Mussolini and wrote that he was 'a man who . . . has shown a driving force almost unparalleled in running a country . . . one felt that if faced with a difficulty, he would get out his sledgehammer and strike at its root'.

Details of Deterding's financial support of Mussolini remain secret but it probably took the form of the help he extended to Franco in Spain. This was publicly exposed by Hugh Dalton, an academic economist, Labour frontbencher and future Chancellor of the Exchequer who made what was reported as 'an indignant speech' in the Commons about 'the very large credits being made available to Franco' by Deterding and Shell.

There can be little doubt that Deterding had been supporting Hitler from the early 1930s. He was obliged to deny in 1931 that he had offered the Nazis a £30 million 'loan' in return for a petroleum monopoly, and was alleged in the following – German presidential election – year to have proposed a £50 million credit scheme. But it is a matter of Foreign Office record that in 1935 he sought in negotiations to provide Germany with a year's supply of oil – in essence, a military reserve – on credit.

With this revealed, Shell at last felt obliged to blow the whistle. The talks were swiftly terminated and Deterding, at the age of 70, was finally eased out of the door on 17 December 1936. But as though to underline his commitment to the Nazis, Deterding almost immediately divorced Lydia Pavlova and married Charlotte Minna Knaack, his buxom German confidential secretary and 'adviser on German affairs'.

In retirement, Deterding chose to live neither in the land of his birth nor in the country that had honoured him. Instead, Sir Henri took the third Lady Deterding to a country estate he had bought near Dobbin on Germany's Baltic coast. And here, married to an enthusiastic admirer of

Hitler's economic policies, he was at last able to 'feel, think and act like a true Nazi'. Thus was his betrayal of Marcus Samuel – the man who had once said to him 'I am going to pay you the highest possible compliment. You should have been born a Jew' – complete.

Deterding died in February 1939. His funeral was chiefly remarkable for having a Nazi functionary stomping about with wreaths from Field Marshal Göring and the Führer himself, in whose name and on whose instructions Deterding was declared 'the great friend of the Germans'. But death did not – quite – extinguish Deterding's connection with the Third Reich. For when on 29 April 1945 Hitler sent his last despairing message from the bunker, essentially demanding that the sycophantic Field Marshal Keitel work a miracle and save the situation, he sent it to a country house on the Baltic coast. Keitel had set himself up in Henri Deterding's last home.

CHAPTER 5

IN THE SHADOW OF GIANTS

CAPITAL CONCERNS

The company that Henri Deterding was finally obliged to quit was, during the Second World War, at the heart of the British government's Petroleum Board. Better known as 'the Pool', the organisation lumped together almost 100 petroleum distributors and, with all competition suspended for the duration, ran the entire oil import, storage and distribution system.

The Board was chaired by Shell's Sir Andrew Agnew, the director identified in *The Prize*, Daniel Yergin's superb history of the oil industry, as being chiefly responsible for blowing the whistle on Deterding's attempts at supplying oil to Hitler's Nazi regime. With wide powers, excellent telephone and teleprinter communications and a substantial workforce which included 6,000 tanker drivers, the Pool controlled road, rail and pipeline oil transport systems, and was reckoned to be an organisational triumph. The importance of the Board – as its chairman, Agnew had a seat on the War Cabinet's Oil Control Committee – underscored the way in which petroleum had in half a century moved from the periphery to the heart of national life and affairs.

Fiercely regulated within three weeks of the declaration of war in September 1939, petrol was one of the first commodities to be rationed. But pleasure motoring was not initially banned even though many car owners complained that a month's ration could be easily exhausted in less than a day.

It was soon after the shocking news from the Far East of the sinking within an hour of the British battleships *Repulse* and *Prince of Wales*, and the fall of Singapore with the surrender of 70,000 Commonwealth troops, that pleasure driving was outlawed in March 1942. The figures tell the story. Private motorists were allowed 823,000 tons of petrol in 1940. Two years later, the amount available to essential personnel and key workers permitted to use private cars in their work had fallen by almost 50 per cent to 473,000 tons. In 1943, the ration was reduced still further to a total of 301,000 tons.

If the war had greatly reduced sales of petrol to private motorists, it spurred the invention, development and manufacturing of chemicals on a previously unimagined scale. Shell only became seriously interested in the chemical business in the late 1920s, and then mostly at the urging of Robert Waley-Cohen, who had, of course, read chemistry at Cambridge.

A wartime triumph for the company's chemists was the invention and development in fewer than 40 days of a waterproofing treatment that enabled 150,000 vehicles to be landed in the surf and driven up the D-Day invasion beaches without 'drowning'. Commissioned by the Ministry of Supply in January 1943, Shell produced more than 10,000 tons of the easily applied treatment.

Walter Samuel proved supportive as chairman of this rapidly expanding sector of the business and told shareholders he had no doubt that chemicals would play an increasingly important part in Shell's peacetime activities.

The second Viscount Bearsted – Walter Samuel had of course inherited his father's title – remained chairman of Shell during the first year of peace. But his health was beginning to fail and in early 1946, at the age of 64, he told fellow directors that he wanted to step down. Walter Samuel, who had been on the board of the company since 1907 and was only its second chairman in 49 years, went in July 1946. His successor was Frederick Godber, a Shell 'lifer' who had been knighted in 1942 having begun his career as one of Henri Deterding's Billiter Street office boys. As well as once being fired by Deterding – something Godber ignored and the Dutchman had forgotten within hours – the new Shell chief was fond of recalling that he had also been warned by a finger-wagging Marcus Samuel that untidy handwriting would seriously jeopardise his career prospects.

Godber, who was made a peer in 1956, used to claim that he had educated himself in business matters by scrutinising every incoming letter. He would remain Shell's chairman for 15 years and his leadership would be tested in a very different – and rapidly changing – post-war world. Moreover, Godber's task was made more difficult still because he was obliged to operate in the shadow of giants. With Marcus Samuel and Henri Deterding, everything about the management of Shell was up close and personal; you would not have gone to either man for a coolly objective, dispassionate view of any major corporate matter. Walter Samuel had in reality made little difference at the top of Shell. He might have been chairman but everybody knew that Deterding carried the big stick and called all the shots. Godber was actually taking over from men whose idea of running a company was to dominate it utterly.

An anti-communist ideologue in the familiar Shell pattern set by Marcus Samuel, Henri Deterding and Walter Samuel, Godber was obliged to watch the march of Marxism across Eastern and Central Europe, China and North Korea. He was also little more than an angry spectator when the company's operations in Czechoslovakia, Yugoslavia, Hungary and Romania were either nationalised or forced into dissolution.

There were anyway bigger and more pressing problems on Godber's agenda. Shell had emerged from the war in a badly beaten-up condition. The company's production, storage and transport facilities had taken a pounding all over the world from Western Europe to the Far East.

The physical restoration of war-wrecked facilities would alone prove massively expensive. But Godber also knew that if Shell wanted to stay in the big game, rapid and costly expansion would be essential to grab and retain a significant share of the huge peacetime market for everything from petrol to the new, oil-based materials synthesised by chemists in their wartime laboratories. Godber thus judged Shell to be in urgent need of large volumes of money and set about the immediate doubling of the company's capitalisation from (in 1947) £43 million to £88 million.

He ran into instant and serious difficulties: Shell's size, strategic significance and economic importance made it essential – indeed, mandatory – to consult with governments wrestling with the problems of war-ravaged economies before setting in concrete decisions affecting the company's long-term future. And it was now that Godber and his fellow directors discovered just what a handicap Shell's structure could be in achieving rational progress. For the cash-strapped Dutch authorities, confronted by the need to find in their money markets the fat end of a serious investment bill in the 60:40 ratio attending all such Royal Dutch/Shell matters, were not at all happy.

And they were even less happy when it was pointed out that the only cheaper alternative would be to reduce Royal Dutch liability by restructuring the company in a genuine 50:50 arrangement. A constantly recurring theme in Shell's history, the prospect of a real merger had last been discussed in 1942 when, because of the German occupation of the Netherlands, the Dutch government was sitting in London.

On top of all these difficulties was piled the limit imposed by Britain's post-war Labour government on the amount of capital private business and industry could raise for investment. And despite Marcus Samuel's cherished dream of Shell becoming an integral part of the great imperial machine – to say nothing of the pomp and circumstance with which it sometimes behaved at home and abroad – Shell was in official eyes at one with the corner shop in remaining an entirely private enterprise. Moreover, with gaping, bomb-blasted holes in the housing and infrastructure of almost every major British city, huge amounts of money were being channelled by the government into reconstruction nationwide.

In the end, after much haggling and the involvement of both the British and Dutch governments, Godber had to settle for what he called 'the low figure of £10 million' in increasing Shell's capital to £53 million. He made plain the fact that he was far from pleased with the outcome and laid the blame 'solely with the Dutch government who just dug their toes in'.

The issue caused considerable friction between the British and Dutch boards of the company. John Loudon, a man of great ability and no less charm who would later be described as 'the true father of Shell in the last third of the twentieth century', said the matter 'was not always diplomatically handled' by some of the most senior people involved. Another director described the row as 'destructive nationality-wise'.

The irony was that within five years and by way of considerably less fraught means, Shell's capital was not only increased to the £88 million

Godber sought but also without any disturbance to the 60:40 arrangement.

CRUDE HUNGER

There was, however, no time to dwell on the finer points of intra-company diplomacy because another recurring corporate theme – the craving for crude – was demanding serious attention.

Shell has for many decades been viewed by the other oil giants and by industry analysts as a 'crude hungry' company. From the mid-1940s, the inability to satisfy its own appetite could in part at least be explained by the destruction of its Netherlands East Indies sources. Later, the lack of a major onshore Middle Eastern source would be held responsible.

In truth, Shell has always seemed more interested in, and certainly better at, transporting and marketing oil than finding and producing crude in volume. Marcus Samuel had, after all, begun the business by shipping and marketing Bnito's Russian oil in the Far East and later, in two deals which amply demonstrated the dangers of reliance on 'outsourced' supplies, went first and disastrously to Moeara Enim and then catastrophically to Guffey and Spindletop for much of Shell's oil.

And yet, many years and two world wars later, it was to Gulf Oil – a direct corporate descendant of Guffey's crashed enterprise – that Godber turned in 1948. For although Shell had an 11.5 per cent share of the global market and was producing slightly more than 750,000 bpd, the company was again in the all too familiar position of being short of crude. Shell had the capacity to refine and sell vastly more oil than it could get out of the ground.

Gulf, by contrast, was fairly bulging with crude that it could neither process nor market. The two companies thus rapidly came to terms under which Gulf supplied crude to Shell for refining and shipping to customers. Once costs and expenses had been taken into account, the profits – and they were considerable – were shared 50:50.

Brokered to last ten years, the arrangement worked so well that the partners kept the deal running for almost a quarter of a century. There can be no doubt that the arrangement enabled both companies to benefit from the boom triggered by the release of consumer demand kept dammed by rationing and shortages during the war.

In the United States, oil consumption rocketed, tripling from 5.8 million bpd in 1948 to 16.4 million bpd in 1972. Car sales, supercharged by the provision of cheap credit facilities to demobilised servicemen, raced ahead so that by 1950 there were 45 million private vehicles on the road, a 60 per cent increase over 1945. Demand for oil-based synthetics used in everything from fashion goods to fishing lines leapt to dizzy heights.

Even in battered Britain, where rationing of some commodities remained a fact of life for years, the savage winter of 1947 – one of the longest and coldest on record with snow 20 feet deep in some rural areas – resulted in greatly increased demand for heating oil. As coal became increasingly expensive across Europe, substantially wiping out its traditional low-cost advantages, oil was seen as a convenient, less messy and more cost-efficient alternative.

But the biggest and most profound changes to the petroleum industry and the way it worked were now under way in the oil-producing countries.

Ever since the death in 1935 of the detested dictator Gómez, Venezuelan governments had been under pressure to use the nation's oil wealth to improve welfare in general and the wretched condition of the poor in particular. Contracts between the Caracas government and Shell and Standard of New Jersey, the two majors responsible for the production and refining of most of Venezuela's oil, had for more than a decade been giving the authorities more of a participatory role in the industry. This was not, of course, the result of an inexplicable attack of corporate altruism: fears of expropriation were in Central and South America never far from negotiators' minds. Nevertheless, in 1948, genuine 50:50 contracts were entered into.

The impact on the global industry was huge and immediate. Within four years, every Middle Eastern producing country – with the single, mini-state exception of Bahrain – had made a similar deal.

The extraordinary new powers conferred by the possession of oil, first revealed by the Venezuelan deal and swiftly underlined by producing countries in the Arab world, were further exemplified by the confused drama about to unfold in Tehran and Abadan.

Iran in the immediate post-war era was the world's third largest producer of crude and, in Abadan, had the world's largest refinery. Most of Iran's oil was produced by Anglo-Iranian, the company originally named Anglo-Persian, in which, as will be described later, the British government, so anxious to have a secure supply with which to fuel the fleet, had a majority (51 per cent) shareholding. But Anglo-Iranian had operations elsewhere and had, over the years, achieved the status of a major player.

Under the terms of a 1933 agreement, Iran not only received royalties on its own oil but also a 20 per cent share of Anglo-Iranian's profits. This was thought by the company's chairman, Sir William Fraser, to be an arrangement so generous as to be suggestive of weak-mindedness. And Fraser, loathed and feared in roughly equal measure by those who worked with him and, it must be said, also by a great many who did not, hated weakness in any form.

In truth, Fraser seemed utterly unconcerned that at a time when the oil industry was being turned upside down by the new 50:50 agreements, his company and the British government were together making three times more from Iranian oil in profits and tax revenues than the Tehran authorities were receiving in royalties.

Known throughout the industry as a fearsome negotiator, Fraser was uniquely unsuited by nature and temperament to run a government-owned operation in crisis and at a time when silky diplomatic skills of a high order would be essential. Fraser, regarded by Whitehall mandarins as a coarse-grained and mulishly obdurate autocrat, despised politicians and civil servants whose questions and observations he dismissed with contempt as the unwarranted interference of know-nothings. But even Fraser in the dying days of the 1940s had come to sense that he and his company were in serious trouble. Escalating Cold War tensions, ratcheted up during the Berlin Airlift, would soon be turned into a major military conflict by North Korea's invasion of the South. In Iran, the communist-led Tudeh Party was gaining increasing support and there had been clashes on the border between Soviet and Iranian troops. Moreover, the clamour for a better – Venezuelan-style – deal for Iran from its oil was being overtaken by raucous demands for the wholesale nationalisation of the industry.

Fraser was sufficiently moved by events to offer the Iranians concessions, in the form of a modification to the 1933 agreement which would have given them increased royalties and a large one-off cash payment. The government accepted the new terms but they were not put to the Iranian parliament for nearly a year because of fears – wholly warranted – that the opposition would tear them to shreds. When finally the agreement was submitted for scrutiny, there was fury. And in amongst the howls of protest were demands for the deal to be rejected and the oil industry nationalised forthwith.

The opposition to the measure was led by the eccentric chairman of the parliamentary Oil Committee, the veteran radical Mohammed Mossadegh, a lawyer with a flair for histrionics and a penchant for doing business in pyjamas while propped up in bed.

The United States, now deeply concerned about the deteriorating military situation in Korea and the severe strain being imposed on its domestic oil supply, urged the British government to pressure Fraser into making the Iranians a speedily acceptable offer. Fraser, to the surprise of nobody who knew him, dug his heels in. The British he argued with; the Americans he simply ignored.

Mossadegh, however, could not be ignored. In fact, Anglo-Iranian's worst nightmare became reality when in April 1951 the Iranian parliament elected the old firebrand Prime Minister. Fraser now offered the Iranians a convoluted 50:50 deal but it was, of course, too late because Mossadegh had been made premier specifically to nationalise Anglo-Iranian. The company was officially consigned to history on 1 May.

There followed intense bouts of negotiation between Mossadegh, the British and the Americans, and at one point, after an extraordinary 80 hours of talks, it seemed that US diplomats had come up with a winning formula. Stressing the Netherlands connections of Royal Dutch/Shell, they proposed that the company should run the Abadan refinery with Anglo-Iranian buying the oil on a 50:50 basis.

A DAY OF HATRED

Mossadegh agreed subject to a single proviso – that not one Briton should be involved in the running of the plant. There had, after all, just been in Iran a special 'Day of hatred against the British Government' national holiday. Organised by the politically powerful Ayatollah Kashani, it had proved hugely popular.

In the nation's oilfields, production had plunged from 660,000 to barely 20,000 bpd, owing mainly to the effectiveness of a British embargo enforced by the Royal Navy. In fact, significant military action against Iran had been discussed in Cabinet, with Defence Minister Emmanuel Shinwell voicing prescient fears that if Tehran was permitted to 'get away with it', other Middle Eastern governments – such as

Egypt's – might be encouraged 'to try it on'. Intervention by the armed forces was, however, ruled out.

In October 1951, a general election in Britain had seen Labour swept from power and the Conservatives, with Winston Churchill as Prime Minister and Anthony Eden as Foreign Secretary, returned to office. Eden, a polished old-Etonian officer and gentleman with fading 'matinée idol' looks and charm in abundance, was the antithesis of Fraser. He might, indeed, have been purpose-designed for dealing with the Iranian problem given that at Oxford he had read oriental languages and was a Persian specialist.

When Mossadegh's 'no Britons' reaction to the American proposal was reported to Eden, he was outraged and dismissed the deal out of hand. Mossadegh now gave the remaining British technicians and their dependants in Abadan seven days to quit Iran. On the morning of 4 October 1951, a British gunboat was sent to Iran. The cruiser *Mauritius* sailed in to take the oilmen and their families to Basra, and Britain waved a humiliating goodbye to its biggest overseas asset.

But that was by no means the end of the affair. For Iran's oil was needed in the market – the shortfall caused by the naval blockade had been largely made up by increases in production elsewhere in the Middle East – and America and Britain planned and executed a coup aimed at toppling Mossadegh and putting Iran firmly under the control of the Shah.

Called Operation Ajax by its CIA and MI6 perpetrators, the coup would, it was hoped, capitalise on a wave of discontent caused by the hard times now afflicting millions of Iranians because of the country's inability to sell its oil. Moreover, Mossadegh was also causing America – and, to a lesser extent, Britain – grief through his increasingly cosy relationship with the Soviets.

Staged in August 1953, Operation Ajax miscarried in almost every respect and came perilously close to total failure. The signs had been comprehensively misread and misinterpreted. To be sure, many Iranians were resentful of being plunged into poverty by the collapse in oil sales. But they were no better disposed towards Britain than they were when they had wholeheartedly supported demands for the nationalisation of Anglo-Iranian. In truth, to most Iranians the link between their diminished incomes and the Royal Navy's ships enforcing the embargo was all too painfully obvious.

At one stage, indeed, the coup had misfired to such an extent that it was the Shah who felt obliged to flee Iran with his family to escape the wrath of Mossadegh's enraged supporters. By and by, the situation was reversed and at the end of the month the Shah was again on the throne while Mossadegh languished in jail.

That, however, still left unresolved the central question of what to do about Anglo-Iranian. Rab Butler, Britain's sometimes splendidly indiscreet Chancellor of the Exchequer, candidly admitted that he and his Cabinet colleagues were totally 'stumped' by the issue. In the end, after much agonised argument, the Americans proposed that a consortium of major companies should produce and refine Iran's oil. But, as so often happened with this vexed issue, even this anodyne idea touched off a bitter row and what amounted to a turf war between Washington's State and Justice Departments.

With all the enthusiasm of men invited to do business inside an active volcano, the representatives of the seven biggest oil companies finally agreed to buy 60 per cent of Anglo-Iranian's interests in Iran for $90 million plus royalties to an additional maximum total of $510 million. Shell's laconic John Loudon signed up for 14 per cent of what was on offer. It was, he observed, 'a wonderful deal for Fraser' given that Anglo-Iranian had already been nationalised and in reality had nothing to sell to anybody.

A FORECAST CRISIS

And it was Loudon who played a key role for Shell in the even more tangled, fraught and, for Britain, politically catastrophic Suez affair.

For almost 75 years, Egypt – and its crucially important Suez Canal – had effectively been under British control. What began with invasion and developed into rule by military occupation had in more recent times given way to the somewhat subtler manipulation of successive client regimes.

In the early 1950s, however, Egypt – like Iran – was in ferment. A coup organised and executed by military officers had in 1952 swept King Farouk, a young and prodigiously fat voluptuary, out of power and into exile. But only two years later, the coup leader, General Mohammed Neguib, was toppled by Colonel Gamal Abdul Nasser, a much younger and even more charismatic Arab version of Iran's Mossadegh.

Nasser, like Mossadegh, was a beguiling speaker who used the powerful Voice of the Arabs radio station to propound his heady vision of a new, Pan-Arab world in which dignity would be restored to a people who had suffered the hurt and humiliation of the flight from Palestine and the creation of the state of Israel on their land. Only collective action could bring hope to the despairing Arab masses reduced to serfdom by their rapaciously exploitative colonial masters.

The prototype for a score of young nationalist militants who would rise to power in the Middle East, Asia and Africa, Nasser had the gift of speaking directly to his listeners' hearts and minds. He could, and very often did, bring crowds of tens of thousands onto the streets to protest against the supine behaviour of Arab governments and their bloated bureaucracies in the face of challenge.

On the home front, Nasser wanted immediate action on the Suez Canal. Egypt was a populous nation crippled by poverty. Oil-producing states, Nasser pointed out, were now receiving 50 per cent of all revenues. But the major share of the Suez Canal Company's earnings – derived, of course, from tolls paid mainly by oil companies such as Shell – was still going to foreign shareholders. And the biggest of these and thus the major beneficiary was the British government. The parallels with Iran and Anglo-Iranian were as uncomfortable as they were obvious.

Shell, whose foundation had depended on Marcus Samuel's creation of a fleet of tankers designed – by Fortescue Flannery – specifically to carry oil in bulk safely through the Canal, had been monitoring events in Egypt closely and well.

Nasser nationalised the Suez Canal in July 1956. There was outrage in Britain, France and Israel. All were apparently surprised, not to say stunned, by the Egyptian leader's timing. Shell, a weary veteran of expropriations, nationalisations and sequestrations, had by contrast anticipated and indeed forecast the seizure of the Canal three years before it happened. Moreover, the company's analysts – whose track record in political prognostication, including the collapse of communism and of the Soviet regime, remains unparalleled by either the State Department or the Foreign Office – had devised a scheme aimed at forestalling the move.

In the period between the toppling of Farouk and the taking of power by Nasser, John Loudon put Shell's plan to the French president of the Suez Canal Company, whose operating licence would expire in 1968.

Loudon said that the key to the scheme was the return of the Canal to Egypt by the Canal Company, who would then operate the waterway on leaseback. This would meet the demands of the nationalists, provide a measure of security for Canal users and a long-term future for the Canal Company.

The idea, however, proved far too revolutionary for a company whose management, according to a visiting American politician, was not so much conservative as 'fusty, hidebound and nineteenth century in style'.

In the wake of nationalisation came three months of furious diplomacy resulting in little more than the setting of the stage for disaster. For on 29 October, and precisely as had been agreed with Britain and France at secret talks in Sèvres five days earlier, Israel launched an attack on Sinai.

The Anglo-French position was based on the certainty of the Israeli action triggering a vigorous military response from Egypt. With battle joined, Britain and France would then issue an ultimatum on the pretext of 'protecting' the Canal. If fighting continued, as was almost inevitable, British and French forces would together invade the Canal Zone with the aim of overthrowing Nasser and securing a new Canal settlement.

Everything about the operation had been misconceived and miscalculated on a barely credible scale. The United States, which within days would be going to the polls to elect either Eisenhower or Stevenson as president, was wholly opposed to the venture. It had in Washington's eyes the look of old-style and deeply unattractive colonial bullying with a built-in capability for triggering something much bigger and infinitely more dangerous.

By 5 November, the Israelis, who had their own reasons for attacking Egypt, had sewn up Sinai and the Gaza Strip. British and French forces now began an airborne assault – exactly 24 hours after Soviet tanks and troops had thundered back into Budapest to crush, brutally and bloodily, the Hungarian uprising. Any hope of a coordinated, coherent Western response from the moral high ground to Moscow's murderous action had perished in a welter of machine-gun fire on the banks of the Canal.

Thus was catastrophe wrought out of what a little earlier had been merely disastrous. Britain and France were condemned by almost everybody in the UN. The Soviets, seemingly without a scintilla of embarrassment at their hypocrisy, not only denounced the allies but also threatened military intervention on the side of the Egyptians. And on 6

November, London and Paris now heard that a furious President Eisenhower, elected by a landslide, was considering oil sanctions against them.

Britain's humiliation was complete. A ceasefire was called and soon the troops were on their way home. The Suez affair, which had riven political parties, divided families, caused violent arguments in cinema and fish shop queues, and sparked protests and even fist fights in normally placid British towns, had been the last roll of the imperial dice. Britain, clearly, was no longer a great power; without the backing of the United States, she was no kind of power at all.

Prime Minister Eden, who as a result of botched surgery had been life-threateningly ill shortly before the military action began, was obliged to resign. He left Britain to recuperate in the West Indies with the hollow-cheeked, haggard look of a broken and haunted man who, often in severe pain, had been fed a nightmarish combination of powerful analgesics and amphetamines.

With the Canal shut – block ships had been sunk in the waterway during the fighting – and much of the West on the brink of an energy crisis, the Oil Lift, a programme of cooperative action among governments and oil companies in the US and Europe, swung into action. Oil supplies and tankers were shared and vessels were comprehensively re-routed to make maximum and most efficient use of all facilities.

Petrol was again rationed in Britain within 45 days of the Canal being blocked. Not only were motorists limited to only enough fuel for about 200 miles per month, they also had to pay a third more for it: a 65 per cent increase in government tax and much higher shipping and transport costs were blamed.

The Suez Canal reopened to tanker traffic in April 1957 and a month later the British government, with little evident enthusiasm, directed shipping to use the waterway again. At the same time, petrol rationing was ended in the UK. Diplomatic relations between London and Cairo were, however, not restored until December 1959, by which time Shell had been back in business in Egypt for more than a year.

The company, like all the oil majors, had learned some important lessons from the crisis. Bigger – and soon, vastly bigger – tankers were an obvious and inevitable consequence of the Canal's closure. Japanese marine architects and shipbuilders were able to bring a new generation of steels, welding technology and greatly improved diesel engines together in vessels which grew from 35,000 tons through 110,000 to 200,000 tons remarkably quickly.

Yet again in oil, size demonstrably mattered. The new maritime giants, ploughing through the oceans at a steady 14 knots laden with monster cargoes of crude, changed the economics of the transport business for all time. The Suez Canal, in the grip of another major international crisis, would in the future again be closed, only this time for eight years rather than a matter of months. By then, 200,000-ton tankers would seem merely average and many Western consumers, bothered only about being able to fill their cars with petrol and their boilers with heating oil in winter, would neither know nor care whether the Canal was open or even if it still existed.

Securing supply exercised Shell greatly in the wake of Suez and forced changes in refining strategy as well as spurring exploration efforts almost everywhere from Africa to New Zealand. There was good news on this front, too. Shell had been looking for oil in Nigeria since 1938 and found it onshore in 1953. Following intensive development work in the Delta region, the country was set to become one of the world's largest oil producers and of critical importance to Shell.

But the Nigerian operation would also turn into a long-running PR catastrophe for the company by feeding the world's print and broadcast media with stories and images of brutal violence, horrendous pollution, massive corruption and, most damagingly of all, the execution of the writer and environmental activist Ken Saro-Wiwa.

The mess that Shell got into – and, by its maladroit handling, made infinitely worse – in Nigeria has baffled many industry observers. For the company in the post-war era, and chiefly as the result of the ideas and policies of John Loudon, had won the reputation of being the best informed, most internationally minded and diplomatically adept of the majors.

The respected oil industry historian and commentator Anthony Sampson, in his *Anatomy of Britain* and *The Seven Sisters* books, emphasises that Shell in the '50s and '60s 'pushed through *ahead* of politics' the recruitment and rapid promotion of local employees.

One result was the growth of the company into a huge and truly multinational organisation employing about 200,000 people in more than 120 countries and, by the mid-1960s, earning a staggering £3.5

billion a year. Another was the swift and not entirely painless death of the imperial ethos that Marcus Samuel had bequeathed Shell.

But like the United Nations, which it had come to resemble in more ways than one, Shell had also become an administrative nightmare of Byzantine complexity: big had become huge and, in management terms, anything but beautiful.

The internal changes Shell underwent in the immediate post-Deterding period produced at least as many problems as they solved. Given the firm's bizarre structure, the possibilities for muddle, duplication and the compounding of confusion into chaos were limitless. The company had even managed to evolve its own – much satirised and ridiculed – Shellese language in which the opaque was always preferred to the clear, and the otiose to the simply essential.

WRESTLING WITH SERPENTS

In 1959, a serious attempt was made to get to grips with the problems and, in what was regarded as a breathtakingly bold move, the American management consultants McKinsey & Company were brought in – largely at John Loudon's insistence – to review Shell's organisation.

Loudon was part of the oil aristocracy. His father had been chairman of Royal Dutch, his grandfather had been a governor general of the Dutch East Indies and his uncle was foreign minister. A Jonkeer – the Dutch equivalent of a baronet – as well as a KBE, Loudon spoke five languages fluently and had the polished manner of a senior diplomat.

Tall, wavy-haired, clever, courteous, urbane and possessed of an instinctive flair for public relations, Loudon's value to Shell was vastly more than ornamental. In 1944, at the age of 39, he was sent to run the company's Venezuelan operation at a time of particular sensitivity when the first 50:50 agreement was in the making. As well as working on this, Loudon also instigated a rapid and successful programme of Venezuelisation.

As head of Shell's committee of managing directors, Loudon helped McKinsey's men wrestle with the serpentine ramifications of the company's crazy structure and organisation. Nine months later, McKinsey produced a largely secret report, nearly all of whose major recommendations were accepted. Given that most of these were reckoned to be Loudon's ideas, this was not entirely surprising.

The biggest exercise of its kind ever commissioned and undertaken in

Britain, the McKinsey report produced some worthwhile results. A new London-based outfit, Shell International Chemical Company, was created to handle the hugely expanded chemical business. London and The Hague were brought under the same organisational umbrella, some senior men were quietly put out to grass, administrative staff numbers were reduced and the seven wise men at the pinnacle of the pyramid were told to devote more time to thought and less to day-to-day matters.

Down the line, the changes were not greeted with universal rejoicing. Some middle managers thought that 'streamlining' had succeeded only in the creation of further layers of complication. They were not even mildly surprised when four years later, with many of the old problems still self-evidently alive and well, Loudon called for a review of the McKinsey-inspired reforms in the hope, he said, of making 'further improvements'.

In truth, Loudon's legacy was a company which, if far from organisationally perfect – or even coherent – was sound in mind and limb, generally well thought of and sufficiently robust to withstand the firestorms and earthquakes generated by the OPEC challenges, oil prices rocketing on the spot market to $40 per barrel before crashing to $6 and less, Middle Eastern wars, revolutions in Libya and Iran, and a ferocious row with Prime Minister Ted Heath in Britain.

And long after he had retired, there was still something Loudonesque in tone about Shell documents such as *Profits and Principles*, published in 1998, an attempt at demonstrating that there was no inherent conflict between the getting of oil, the maintenance of a decent environment, an informed interest in and concern for community issues and the making by oil companies of handsome profits. Shareholders, it is said, liked the document. As we shall see, however, it played less well in the Niger Delta, among people living near the Motiva refinery in Texas and conservationists counting the dwindling number of grey whales off Sakhalin Island.

But no amount of organisational tinkering or sophisticated PR pamphleteering could disguise the fact that Shell had direct control of barely more than 20 per cent of the Middle Eastern oil it was handling. More than ever now, it was a company ravenous for crude.

Driven by hunger, Shell became increasingly prepared to take on serious technical challenges to develop relatively small accumulations of oil locked up in difficult geological conditions. Making economic sense of some of these high-cost fields – of limited or zero interest to the other majors – obliged the company to develop what it calls enhanced oil recovery techniques designed to squeeze as near as possible the last drop from the rock.

CATHEDRALS IN THE SEA

But the greatest technical and financial challenges were on the company's own doorstep in the North Sea.

The existence of oil and gas beneath the waves was determined in the mid-1950s. Shell and Esso, in a joint venture, in 1959 provided an important clue to location with the discovery of the massive Groningen gasfield in Holland.

Besides its importance as the biggest gas find – the equivalent of more than 1 billion tons of coal – in mainland Europe outside of Russia, Groningen's closeness to the North Sea and the extension offshore of remarkably similar geology was of serious interest. But it was not until ten years later – and after Shell and Esso had found gas in the southern North Sea in the Leman field – that Phillips Petroleum made a major discovery of top-quality light oil in Ekofisk.

At the end of 1970, BP made a significant find in the Forties followed in rapid succession by big strikes for Shell and Exxon in what would become the mighty Brent field.

Finding abundant gas and oil in the North Sea was one thing; getting it out from deep beneath the seabed and into onshore processing plants was quite another. The problems and the financial risks were prodigious.

The depth of the North Sea was not a serious difficulty. The weather was. Shell reckoned from the outset that drilling and production platforms would have to be capable of withstanding waves 100 feet high driven by winds of up to 160 mph. In addition, there were powerful currents and tides to contend with as well as the usual ocean hazards such as busy shipping routes and fishing grounds, plus a long storm season and thick fogs.

It was in the design and construction of the platforms – each big enough to dwarf a European cathedral – that Shell and the other companies involved in the North Sea began engineering on a truly heroic scale. This was, in every sense, big stuff that Brunel would surely have understood and certainly applauded.

Some idea of what was entailed can be gained from *The Petroleum Handbook*, a sober, 700-page A–Z of oil, not noted for its lightness of touch or inflammatory language, produced by Shell for senior staff:

> On land, the wellheads, pipework, power and process equipment of an oilfield may be laid out over 25 square kilometres of ground. Cramming that hardware onto a single offshore platform with a deck area half the size of a football pitch . . . has called for new thinking.
>
> One such platform may have to cope simultaneously with the drilling of 30 or even 40 wells . . . the production of oil from several of these wells; the treatment of the oil to separate gas and water; the treatment of seawater for injection; the re-injection of gas at pressures of up to 6,000 psi (415 bars); the delivery of gas and oil by pumping to the shore; the generation of 14MW of electricity (enough to light a small town) to power all the systems.
>
> In addition, up to 200 men have to be housed, fed and even entertained in their off-duty periods, supplies have to be lifted from boats and helicopters must land and take off both men and equipment.

The costs of operating in the North Sea proved to be astronomical. The *Handbook* states:

> The Brent Field alone, with its four platforms and its separate oil and gas pipelines to the shore, has cost more than £3.5 billion.
>
> It may cost £5 million to drill just one exploration well, £350 to fly one man to his offshore work location and back and £50 to ship one tonne of cargo from shore to platform.

The business of getting men to and from platforms far out at sea proved so expensive and time consuming that a special and very large helicopter, capable of carrying 44 men from Aberdeen the 480 km to the Brent, Dunlin and Cormorant fields in less than two and a half hours, had to be developed. It goes without saying that the machine also had to be capable of doing this in the foulest of weather.

When development work was at its peak in the offshore fields, so many men had to be housed, fed and watered in the North Sea that redundant drilling platforms were converted into 'flotels' – accommodation vessels for up to 500 men. These were anchored alongside fixed installations and linked by a gangway so that the men, as Shell explained, 'were housed conveniently close to their work site'.

In bad weather, the gangways were raised and the flotels were hauled back along their anchor chains to avoid any risk of collision. The men were in these conditions ferried between flotel and platform by helicopters flying shuttle services. 'Either way was one hell of a way to go to work in the morning,' a flotel veteran recalls.

The scale of North Sea operations was underlined in a *Handbook* paragraph on air transport.

> Helicopter traffic, including that for other operators' fields under development in the same area, grew to such an extent that it called for a full air traffic control system to ensure safety. This system in one peak summer month handled 22,000 air movements, only 4,000 fewer than London's Heathrow in the same month.

Now, Shell and the other majors, constantly seeking the cheapest and most easily exploited oil, are winding down their North Sea operations. According to the UK Offshore Operators' Association, exploration activity is at an all-time low and many experts believe the companies will leave untapped beneath the seabed up to 10 billion barrels of oil.

But the experience has made a profound and permanent impact on the way big oil finances its operations. For Shell says that until funding had to be found for the monster platforms – each of which took about five years to build and commission – and long sub-sea pipelines costing £800,000 per mile, 'exploration and production activities were, broadly speaking, self-financing. Now, enormous sums of money are required. A loan agreement for a major project will usually involve not just one bank but a consortium of banks.'

And even for Shell in the twenty-first century, projects do not come much bigger than the $10 billion Sakhalin Island oil and gas scheme off Russia's eastern coast. Shell, leading Mitsubishi and Mitsui in the Sakhalin Energy Company, expects to earn billions of dollars exporting gas to the Asia-Pacific region from a field big enough on its own to meet world demand for four years.

But the company's return to the big game in Russia has also been marked by a fumbling, dissembling response to the protests of environmentalists from 93 countries that the project could wipe out the critically endangered Western Pacific grey whales.

Moreover, Shell's plan to run underground an 800-km-long oil pipeline across any number of faults in one of the world's most earthquake-prone regions is reckoned by protesters to be little short of lunacy.

Shell and the environmentalists agree, however, that a consortium of banks is required for project finance on this scale. The protesters are now making the US Export-Import Bank (Ex-IM), the European Bank for Reconstruction and Development and the Japan Bank for International Cooperation, the institutions considering finance for phase two of the scheme, the targets of considerable pressure.

And in the wake of the return to Russia has come Shell's re-engagement with Libya. The move came at the 25 March 2004 signing of a $200 million gas exploration contract with the Libyan National Oil Corporation.

Malcolm Brinded, head of exploration and production, signed the agreement for Shell at a Tripoli ceremony timed to coincide with a groundbreaking visit by British Prime Minister Tony Blair for talks with Libyan leader Muammar Gadaffi. An official travelling with the premier's party said the value of the agreement could rise to more than $1 billion in the long term. Negotiations about specific projects would continue throughout the year, he added.

Shell became actively involved in Libya, producing 300,000 bpd, nearly half a century ago when the country's high-quality, low-sulphur crude made up a quarter of all the oil consumed in Europe. On 1 September 1969, young army officers – all fervent disciples of Egypt's Gamal Abdul Nasser – seized power in a coup. Their leader was a major named Muammar Gadaffi. The new and radical Revolutionary Command Council swept away all the old agreements with companies such as Shell and, over five years, nationalised the oil industry. Shell conducted some exploration in Libya in the late 1980s but left when the country's relations with the West deteriorated sharply.

Today, Libya produces about 1.2 million bpd and has proven reserves of 30 billion barrels. Unlike Shell's, however, Libya's reserves are believed by most experts to be seriously underestimated.

That Shell had obviously been in talks with the Gadaffi government long before there had been any official indication of a London–Tripoli rapprochement came as no surprise to anybody in the oil business. For, as detailed in part two of this book, Shell and the other majors have over the decades developed diplomatic skills of a high order to complement the immense political power they possess. Oilmen will themselves always protest that the contrary is the case and that their companies are but poor, weak, politically neuter vessels blown hither and yon by the great winds generated and directed by the sovereign governments of the territories they operate in. But this is, of course, demonstrably disingenuous nonsense put about by corporate spinners who, in the case of Shell, have created not so much a false picture as a parallel universe in which everything is unreal.

In efforts to maintain 'our good name' – that quality so greatly prized and understood by Marcus Samuel and even Henri Deterding at his most megalomaniacal – the spinners have sought always to preserve Shell as the shining solution when the unvarnished reality is that the company has more often than not been the problem.

If the reserves crisis did nothing else, it helped reveal the spinners' work for what it was – a web of deceit in which Shell itself became trapped.

PART TWO
Shattered Image

CHAPTER 6

WHO'S RUNNING THE COUNTRY ANYWAY?

The last two decades of the twentieth century saw the growing influence over, and in many cases dominance of, government policies by non-elected, unrepresentative corporate business. What was particularly shocking about this development was the manner in which power and influence were surrendered by various administrations, in what can only be considered a betrayal of democratic processes. Nowhere was such a process in greater evidence than in Britain and the United States of America, particularly during the Thatcher–Reagan years. It is a catalogue of surrender which future generations will have good cause to castigate, leading them to condemn those who facilitated such a corporate takeover – one that led to the usurping of the power and responsibility of elected governments – and forfeited the duty invested in them as members of parliamentary government. In short, sharp and brutal terms, it was a *coup d'état* without the tanks, one that, particularly in Britain, has been given a whole new dimension and impetus by the government of New Labour and Tony Blair, who has presided over an expansion of corporate control over the lives of Britons of which Margaret Thatcher only dared to dream.

It is, however, a process which has had an extraordinarily long period of incubation. In a letter to Colonel William Elkins in 1864, a year before the end of America's Civil War, Abraham Lincoln declared in suitably solemn tones, 'I see in the near future a crisis approaching that unnerves me and causes me to tremble for the safety of my country . . .

corporations have been enthroned and an era of corruption in high places will follow.' Later, it fell to another American President, Calvin Coolidge, who had earlier defined the principal occupation and motivation of his fellow countrymen as 'the business of Americans is business', to introduce oil into the equation. 'It is even probable that the supremacy of nations may be determined by the possession of available petroleum and its products.' It is a comment that, today, while merely assuming a footnote in the history of one quite unremarkable American presidency, was prescient of the rise and eventual supremacy of the country's very first oil aristocrat, one who, interestingly enough, embarked upon his career at the time of the Civil War.

J.D. Rockefeller and his Standard Oil were indicative of the rise to power of the big oil corporations, and legislative attempts were made to curb such unbridled influence to ensure fair competition through anti-trust laws. But in an example of corporate realpolitik, the apparent success of curtailing Rockefeller power was, as the twentieth century proceeded on its highly hazardous way, seen all too soon to have been but transitory. In a series of 'counter-revolutionary' moves, referred to by many as 'Rockefeller's revenge', the dominance of the oil industry over America's politics and the American people themselves was re-established. It is a state of affairs in America which, in the opening years of the twenty-first century, given the presidency of George Bush Jnr, who, together with his family and senior members of his administration, enjoys close and enduring links to the oil industry, has prompted a whole new debate on the power of the oil lobby to influence the nation's politics. High on the agenda are questions about the real reasons why the country went to war in Iraq in 2003.

It was during the opening years of the twentieth century that on the other side of the Atlantic in Britain, oil and those who controlled it were giving rise to anxieties of an equally disturbing nature. The age of oil had exposed Britain's vulnerability. With no sources of its own to speak of, an empire that spanned oceans and continents to control, and with a European war of appalling menace looming on the horizon, the fear that Britain could be subject to ultimate political control by the oil giants was giving waking nightmares to the country's leaders. Writing in June of 1914, just two months before the outbreak of the First World War, Winston Churchill commented, 'Look out upon the wide expanse of the oil regions of the world! Two gigantic corporations – one in either hemisphere – stand out predominantly . . .'

Central to the government's concerns was not just a constant need to locate new sources of the commodity but also – as it was more often than not found in areas remote and inaccessible from established lines of transport and communication – the attendant problem of transporting the oil back home once it had been extracted. (Indeed, the unimpeded transport of oil back to Britain was an imperative and is the reason why the country measured oil in tons, the measure used by shippers, as opposed to the Americans who measured it by the barrel.)

It was an increasingly precarious enterprise that, even for a nation state of such power and influence as Britain during the early 1900s, called upon all its abilities and reserves of ingenuity and diplomatic skill. If the Indian Mutiny had shaken British confidence, then the South African Boer War, which came to an end in 1902, had delivered the most salutary reminder that imperial rule from London had, most decidedly, its limitations.

Even given the geographical scope of the British Empire, from far-flung coral strands to high savannah and from scorching deserts to tropical jungle and rainforest, all attempts to locate oil had met with failure. Until, that is, a source was discovered in Burma, today's Myanmar. It was a discovery that, at last, promised to give Britain control over a supply of a commodity now deemed to be essential to its very survival as a manufacturing nation and a great imperial power. And it was, also, a source of supply that the government was determined to deny to Shell, with its foreign background and contacts, contacts that could well make it subject to alien influence. In pursuit of such a policy based on old suspicions, the India Office, which had the governance of Burma within its brief, specifically excluded Shell from being granted any concessions in the field of this new and exciting discovery.

It was an exclusion bitterly resented by the company: by Marcus Samuel in particular, but also by Henri Deterding, whose dominance of Shell had been ensured by his personal policy of pursuing with ruthless intent new concessions wherever and whenever possible. But Shell, and the men who governed it, were never infirm of purpose and their campaign to, in effect, make serious inroads into government influence continued unabated and, indeed, with increasing vigour, particularly when, much to Deterding's subsequent fury, Burmah Oil, the company established by a business syndicate of Scots to develop Burma's oilfields, moved quickly into very healthy levels of profit indeed. But, for Shell, there was worse to come.

In 1910, the government finally made the decision to switch the ships of the Royal Navy from being fuelled by coal to oil – a decision, as we have already seen, that had long been campaigned for by Shell. Yet, the old suspicions of Shell so comprehensively exercised in the corridors of power in Britain were as alive as ever, and indeed had been somewhat reinvigorated by the company's recent merger with Royal Dutch. It was now believed that there were two foreign influences to which the suspect company could fall prey: the giant across the water, America's Standard Oil, a company whose predatory power in the land of its birth appalled many senior figures in Britain's establishment; and, just across the North Sea, the Dutch. This being so, it should not then have come as any great surprise to Shell – which was, of course, well aware of the 'quality' of antipathy being directed towards it – that all the major contracts for the supply of oil to the navy went to Burmah Oil.

Just a year later, however, it looked as though help might be at hand. Winston Churchill, then a thrusting Liberal politician, was in his parliamentary infancy but, ever the master-pragmatist, was constantly on the lookout for high preferment. In 1911, it came his way, with his appointment as First Lord of the Admiralty, a development given a particularly warm welcome by his long-time friend Admiral Fisher.

The Admiral lost little time in using his powers of persuasion on his distinguished colleague, recommending in tones of unbridled enthusiasm the advantages to be gained by establishing a relationship with Shell, particularly commending Marcus Samuel as someone with whom the government should be doing business. 'He is,' Fisher wrote to Churchill, 'a good teapot, though he may be a bad pourer', continuing, in a moment of unusual candour, 'Old Marcus is always offering me a position as a director.'

Churchill took much of Fisher's argument seriously and ordered that an Admiralty Commission of Inquiry be established to look further into the whole question of the official supply of oil. Marcus appeared before the Commission and in a long and passionately expressed submission gave assurances that there was '. . . no alliance, no agreement, no treaty of any kind between ourselves and the Standard Oil Company'. He also took the opportunity to tell the Commission that Britain could be well assured of Shell's total loyalty, an assurance which rang especially true given Marcus's earlier, and much publicised, stated view that Shell's capabilities should be harnessed in the service of the Empire.

While Marcus had impressed the Commission, Churchill remained wary of the co-founder of Shell and became increasingly suspicious of the company, an attitude he stuck to with his usual robust conviction as the price of petrol began to rise, and rise considerably – increases which had commenced shortly after the Commission had completed its work. Indeed, with the car now becoming more readily available to a greater number of people, the price of petrol assumed a political dimension, a point quickly taken up by the motoring public, who, understandably enough, resented such a rise in price at the petrol pumps. There were public protests, London taxi-drivers went on strike and newspapers began to thunder misgivings in their editorials against what they termed 'the petrol ring'.

It was a situation that Shell could not let pass without challenge and Marcus Samuel took the opportunity of an interview with the *Daily Mail* to tell the country and its politicians a few basic home truths. 'The price of an article is exactly what it will fetch,' he is quoted as saying, then continues by reminding the *Mail*'s readers that Shell was in the oil business for profit and that the sale of the commodity would, as is always the case in the marketplace, be decided by supply and demand.

Churchill, at the time very much a Liberal, was, in keeping with his party's principles, mistrustful of monopolies and – not wishing to miss an opportunity to make political capital out of the public mood of resentment at the repeated increases in the price of petrol – made a speech in which he accused the oil monopolies of secret price-rigging. It was, of course, an indirect reference to Shell and an innuendo against the company's business practices.

This infuriated Marcus and, he felt, questioned again the matter of Shell's loyalty to country and Empire. He went on record saying that he regarded Churchill's attack as being tainted with anti-Semitism.

It was, however, the supply of oil, principally the identification of new sources of supply over which the British could exercise some control, which remained the greatest concern. It was a national dilemma to which Churchill personally gave considerable attention, particularly now that the requirements of the navy also needed to be met.

Shortly before Churchill was appointed to high office as First Lord of the Admiralty, the Anglo-Persian Oil Company (which in 1954 was renamed British Petroleum) had been founded, and had now discovered oilfields of considerable promise. Churchill, as a leading Liberal politician and with his thrusting Conservative commitment to unfettered capitalism still some years distant, did not shy for one instant away from a design on Anglo-Persian that would bring it within state control. Indeed, he was, according to a contemporary report, 'in fighting form', which was fortunate given Anglo-Persian's spirited attempts to keep the state well away from exercising any control or influence over its new status as a potential world supplier of crude oil and its highly profitable derivative, petrol.

The designs of the state, championed by Churchill, proved to be overpowering and in May of 1914, just three months before the outbreak of the First World War, it was confirmed to an astonished House of Commons that the government was to have a controlling interest in Anglo-Persian via its purchase of 51 per cent of company shares.

The following month, in a Parliamentary speech shot through with an eclectic arrangement of bombast and vision, Churchill sought to justify the government decision to go into the oil business. He referred, in suitably grave tones, to the danger posed to the country by the power of oil monopolies that shared world markets between them. Then, fully warming to his theme and swept along by the force and pace of his inimitable rhetoric, Churchill delivered a thundering indictment of Shell: one, on this occasion, shorn of his usual appetite for innuendo and which charged that these 'gigantic corporations' could, in effect, hold the consumer to ransom by pursuing a commercial policy that would keep oil and petrol at an artificially high price:

> It is their policy – what is the good of blinking at it – to acquire control of the sources and means of supply, and then to regulate the production and the market price. We have no quarrel with Shell. We have always found them courteous, considerate, ready to oblige, anxious to serve the Admiralty and to promote the interests of the British Navy and the British Empire, but at a price. The only difficulty has been the price. On that point we have, of course, been treated to the full rigour of the game.

Confirming that the government would continue to purchase oil from Shell, even though it now had a controlling interest in the Anglo-Persian Company, Churchill told Parliament that 'we shall not run any risk of getting into the hands of these very good people'. It was a clear confirmation of the government's intention to use its ownership of an oil company to make sure that the oil it purchased from Shell would be fairly priced – obtained at the very best competitive rates it was possible to achieve.

If the company had felt insulted by Churchill's earlier innuendoes suggesting that its loyalties to Britain and its institutions were suspect, and that its commercial instincts would always put profit before patriotism, this latest parliamentary broadside, delivered under the cloak of parliamentary privilege, was, as on previous occasions, especially resented by Marcus Samuel. But all too soon, indeed within eight weeks of Churchill's speech, the national calamity of war fell upon the British and their Empire, and Shell at last had an opportunity to prove well beyond any reasonable doubt the quality and durability of its loyalty to Britain.

Throughout the national trauma of the First World War, the company refrained from overcharging the navy for its oil, and, even given the government's control of Anglo-Persian, remained its principal supplier. It was, however, a self-imposed restraint it could well afford. Indeed, throughout the awesome four-year conflict, Shell made profits as never before, with its ever-increasing trading activities in the Far East proving particularly lucrative.

A BANQUET OF CONSEQUENCES

As the twentieth century continued to unfold, greater grew the realisation of the crucial importance oil was destined to play in the fortunes of whole nations, even if public disquiet was so often in evidence as to the doubtful 'culture' of the industry, which seemed to contaminate all those who came into contact with it. As the Liberal politician and Prime Minister David Lloyd George, a close colleague of Winston Churchill during his Liberal days, said, on 26 March 1920, 'Oil profits generally seem to find their way by some invisible pipeline into private pockets.'

Indeed, there were widespread suspicions regarding the tendency of the industry to have a corrupting effect, given the huge amounts of easy money being generated, money from which certain individuals benefited minus the sweat of their brow. This apparently limitless supply of cash, cash that had the power to buy some and, in effect, blackmail others, particularly those in high office, became a matter of equally high concern. The idea that 'people have their price' and were, therefore, available for purchase, was seen to pose a direct threat to the very principles and conduct of open and honest government.

At the time of the 1973 Arab-imposed oil embargo, the provocatively brave American politician Senator Frank Church was direct in voicing his concern. Church was chairman of a sub-committee, established by the Senate's powerful Foreign Relations committee, charged with investigating the influence exercised by multinational companies on US foreign policy. Speaking in Iowa in December that year, this most unusual American political figure, whose political awareness had been sharpened by his wartime experience in Vietnam and who, consequently, spoke often of the evils of government appetite for secrecy and of his suspicion of multinational companies, turned his forensic attention to the practices of the international oil companies. 'It is time,' declared Senator Church, 'we began the process of demystifying the inner sanctum of this most secret of industries.' Then, with a rallying candour, he continued:

> We Americans must uncover the trail that led the United States into dependency on the Arab sheikhdoms for so much of its oil. Why did our government support and encourage the movement of the huge American-owned oil companies into the Middle East in the first place? We must re-examine the premise that what is good for the oil companies is good for the United States.

It was an approach both brave and controversial, one the Senator continued with his usual vigour when, in his opening address to the Senate's Foreign Relations sub-committee on 30 January 1974, having made it abundantly clear before the hearing began that it was his express intention to principally address the vexed question of the conduct of the world's major oil companies, this son of rural Idaho declared:

> We are dealing with corporate entities which have many of the characteristics of nations, thus it should surprise no one that, when we speak of corporate and government relationships, the language will be that which is appropriate to dealings between sovereigns.

Church, of course, had his detractors and, as the Senate sub-committee hearings continued, the oil companies mounted a robust defence of their conduct and business practices, hiring some of America's best legal minds to put their case. One such legal eagle was the formidable John J. McCloy, who, in a written submission, claimed that the current crisis, the Arab oil embargo, had come about not as a result of the companies practising short-sighted policies and being obsessed with short-term profit, as Church and others charged, but was one born of armed conflict in the Middle East.

But the Senator was not so easily deflected and pursued his point of view that, in the final analysis, the priorities of the oil companies had turned them into hostages of the oil states; that, at the hour of crisis, the companies discovered that they had no leverage at all, which had made the United States itself a hostage to fortune. At the heart of the crisis, Church replied, was the 'singular dominance of the majors', that the United States had encouraged such a concentration and that if this had not been so, the country would not now be in such difficulties.

In his closing remarks to the sub-committee, the Senator identified the nature of the abiding problem, namely that government policy on oil had been kept well concealed from both Congress and the public at large. It was a conclusion with which the Senator's committee concurred, with its published hearing reports clearly stating:

> In a sense, this is the over-riding lesson of the petroleum crisis. In a democracy, important questions of policy with respect to a vital commodity like oil, the life-blood of an industrial society, cannot be left to private companies acting in accord with private interests and a closed circle of government officials.

THE RIPPLE EFFECT

It was a point of view that prevailed on the other side of the Atlantic, where, in the autumn of 1973, the Prime Minister, Edward Heath, was so troubled about the Arab oil embargo, soon to be exacerbated for Britain by a coal miners' strike, that, on 21 October, he summoned Sir Eric Drake of British Petroleum and Frank McFadzean of Shell to Chequers, the country home of British prime ministers. The meeting was so sensitive that it was cloaked in secrecy, a state of affairs that would have been easily breached had anyone been strolling in the garden, given that the Prime Minister, not known at the best of times to speak in hushed tones, when riled was incandescent with rage.

He insisted to the two oil barons that it was their national duty to make sure that the country had adequate supplies of oil, despite the embargo forced upon both companies by those Arab states in which they operated. Both Drake and McFadzean protested, commenting that they were under an obligation to treat all of their customers equally; that, in effect, patriotism came a very poor second to their international commitments. The two men also spoke of their fear that company assets might well be seized by some of the oil-producing states in the Arab world should it become known that, in direct defiance of the embargo they had imposed, oil was being supplied in unreduced quantity to Britain, which, together with the United States, was a prime target of the embargo.

The Prime Minister, now in a volcanic rage, turned his fire with particular ferocity on the BP chairman, insisting that as the government owned 51 per cent of the company, he was principally duty bound to see to it that oil continued to flow Britain's way, without any significant reduction. But Sir Eric Drake had arrived at Chequers well armed with legal advice, having anticipated that Heath would raise such a point. His response was as wily and as clever as one would expect from a man of oil. BP would, he told a red-faced and angry Prime Minister, follow his instructions only on the condition that the government committed them to paper and, what is more, identified, again in writing, those countries that would be penalised by BP in receiving reduced supplies as a result of Britain receiving preferential treatment.

At this stage of the meeting, Edward Heath knew that the game was at an end; that, given all his towering authority, aided and abetted on this occasion by a robust delivery of prime ministerial bluster, the oil companies had prevailed and demonstrated to the leader of an elected government just who the real boss was when the proverbial chips were well and truly down.

But, come the end of the year, the agony and uncertainty inflicted by the embargo was suddenly at an end. The oil siege of Europe by the producing states of the Arab world was lifted, followed in March of 1974 by an end to the boycott of the United States. In Britain, there was particularly good cause for optimism by the end of that year, the fountainhead of which was the growing confidence that North Sea oil reserves would turn the country into a significant exporter of the commodity – Brent Crude as it would become known – and that the Arabs' quadrupling of the price would mean previously undreamt-of revenues flowing into the Treasury's coffers.

Indeed, it had only been the huge hike of oil prices by the Arabs that had made the extraction of oil from the North Sea – a most expensive exercise – financially viable. This new and most welcome turn of events placed Britain's relationship with the oil barons and their respective companies on a completely new footing, one to which the incoming Labour government of Harold Wilson, who had defeated Edward Heath and his Conservative Party for the occupancy of 10 Downing

Street on 28 February 1974, extended a very special welcome.

It was the promise of such high financial returns that put the question of just how control of the oil companies was to be exercised right at the top of the political agenda. A parliamentary report of the previous year, which had highlighted past governments' indulgence of the oil companies, in the form of a whole raft of generous concessions, including the most contentious, tax relief, had concentrated the minds of politicians of all three parties. Even the Conservative Party, traditionally the party of the free market, was now amenable to some nature of closer control over British Petroleum. With a Labour government back in power, it was an exercise that many within its ranks, including ardent socialists such as Tony Benn, approached with particular enthusiasm. It was a situation that the oil companies, BP and Shell in particular, viewed with express distaste and, in some senior circles of both companies, distinct alarm.

They need not have worried, not yet anyway. Indeed, Harold Wilson, for all his perceived 'left-wing' bias, a public perception promoted and nurtured by a predominantly Conservative-supporting press, was an arch-pragmatist and knew well enough the power of the oil barons, with whom he did not wish to tangle. And in any event, with the prospect of the profits to be made from North Sea oil, Britain's long-term ambitions for the sale of oil were not dissimilar now to those entertained by the oil companies, or for that matter OPEC itself, even if, in the short term, this meant higher energy bills for Britain.

As is ever the case when the oil companies put their various strategies into play, plans to counter any government interference in their affairs had been well laid. Indeed, long before the so-called 'wild men' in Wilson's government and within the ranks of the parliamentary Labour Party had begun the tribal chant of taking British Petroleum further into the embrace of the state, the bureaucrats of the Foreign Office were wagging cautionary fingers at Downing Street, warning that any such control would impair the company's standing, and, therefore, the relationships it had built up, in the wider world. Such a stand came as no real surprise to Prime Minister Wilson, who knew well enough that on his retirement as head of the Foreign Office, Sir Denis Greenhill had accepted a place on BP's board. When there are battles to be fought, the oil companies are past masters at having their very own men from high places in, well, high places, both on-hand and on-tap to proffer advice and guidance of the very highest quality.

Yet the principle of control by the state of what was seen, right across the political divide, as a great national resource, was not in any real contention. Indeed, the former Conservative Prime Minister, Edward Heath, had appointed the patrician and highly polished courtier, Lord Peter Carrington, as his Minister for Energy, a man of consummate political skill and experience who had been a tower of strength to Heath during his battle with Shell and BP over oil supplies to Britain at the time of the Arab embargo. And immediately prior to leaving office, in March of 1974, Heath had been even more direct by naming as Minister for Energy Patrick Jenkin, who had put forward the view that the state should be an active 'participant' with the oil companies in the exploitation of such a vital national resource.

The government, therefore, proposed to become a 51 per cent shareholder in all the oil companies which had been licensed by the state to operate in the waters of the North Sea, an association and involvement to be overseen by the establishment of a state organisation, the British National Oil Corporation (BNOC), to be based in Glasgow.

The creation of the BNOC was, in all truth, little more than an extension of Winston Churchill's belief that where such a crucially important commodity as oil was concerned, the state must have a controlling interest.

Such a development enraged the oil companies, with Sir Eric Drake of BP arguing that, yet again, the good faith of the oil companies was being called into question. Shell, while sharing Drake's fury, was not quite so vocal, demonstrating a quality of political skill and deftness that spoke volumes about the company's experience, gained over a period approaching a century, and demonstrating, too, its adroitness at operating for all practical purposes as a 'state' within a state. Shell would, for now at least, bide its time, do, in effect, a Brer Rabbit, 'lie low and say nuffin', or at least very little, as the time was not right by any manner of means to attempt to upturn the government's proverbial apple cart.

But maintaining a dignified silence in the face of what was seen by the oil companies as unjust provocation did not mean, for Shell at least, that it escaped public odium. Even though the Arab-inspired embargo that had placed the oil companies under extreme political pressure was at an end, 1974 proved to be a particularly unsettling time. On 29 May that year, the *New York Times* reported that Shell was to be charged by Japan's Fair Trade Commission with alleged price-fixing. While charges were made by this anti-monopoly body against a total of 12 oil companies,

Shell was the only major company to face such an investigation. And in September, President Gerald Ford delivered a bruising broadside, adding his very own indignation to that of the public over the power of the oil majors and their practice of keeping prices at the petrol pumps artificially high. 'Sovereign nations,' intoned the President of the United States, 'cannot allow their policies to be dictated, or their fate decided, by artificial rigging and distortion of world commodity prices.'

In Britain, some of the oil companies' fears, both real and imaginary, were now to be confirmed. In 1975, Wilson appointed the socialist firebrand Tony Benn as his Secretary of State for Energy, an appointment that sent ice-cold shivers down the backs of both Shell and BP. Benn, even before being handed the Energy portfolio, had made no secret of his belief that North Sea oil should be nationalised and this, allied to his antipathy towards what he saw as the largely unfettered power enjoyed by the oil companies, gave rise to acute anxiety among the oil barons.

Benn, however, was no stranger to practising the charm offensive, as the entry in his diary of 15 July 1975 makes clear:

> At 3.30 Sir Eric Drake of British Petroleum came to see me and I went out of my way to be charming. He said the government holdings of British Petroleum shares must be kept below 50 per cent because it would destroy the credibility of the company in the United States, in New Zealand and elsewhere – BP operates in 80 countries. Therefore, he wanted the BP Burmah shares sold off in the open market but not to foreign governments. Well, I'm not accepting that. I had contemplated giving Drake the chairmanship of BNOC but he was so negative and hostile that I changed my mind. I'm glad I saw him and it was probably a good thing to be on reasonably good terms with him, though he is the most Tory of Tories.

Two days later, in his entry for 17 July, Benn writes of the nature and extent of corruption spawned by oil:

> The papers over the last couple of days have reported the news from New York that Exxon has openly admitted to paying $51 million (£20 million) to Italian politicians and political parties over the last nine years. That's over £2 million a year flowing from a

> single oil company into Italian political funds. The number of people who could be bought, corrupted, suborned, diverted, blackmailed and assassinated with £20 million defies the imagination. One shouldn't be in any doubt at all as to what we're up against and I shall use this if necessary to defend the development of BNOC.

The cancer of wholesale corruption all too present in the body politic of oil, and the corrosive effect this has on individuals and governments, was a stark reality which greatly troubled Benn. He was only too well aware that now Britain was in the oil business writ large, it would mean those charged with dealing with its emissaries, such as himself, would be subject to constant attempts to compromise officially held positions. He was also very much aware that coming a very close second to corruption was the repression practised by many of the oil-producing regimes, an awareness which led him to observe in his diary entry for 10 January 1976:

> One of the most unattractive aspects about politics is that whatever your own aspirations may be, you have to work with others who are repressive. That is certainly true of the Soviet Union . . . and our record in bolstering up the Sultan of Oman's corrupt regime is quite appalling.

Suddenly all efforts at charm were abandoned and, as this particular entry makes abundantly plain, the Battle of the North Sea had most certainly commenced in this British Minister of the Crown's mind. Indeed, once again, the men of oil were on charge. The battle lines were drawn.

Britain's Minister for Energy did not have to wait too long before taking up his ever-ready cudgels. The entry in his diary for 25 January 1976 sees him at Chequers, in the company of the Prime Minister Harold Wilson; David Steel, who had taken over as BP's chairman at the retirement of Sir Eric Drake; Steel's deputy chairman, Monty Pennell; and Frank Kearton, chairman of BNOC.

> Harold asked endless questions about Alaska, Iran and Canada. David Steel, of BP, was uneasy because he didn't really know the figures. Monty Pennell, the Deputy Chairman of BP, was full of facts and Frank Kearton didn't say a word . . . David Steel then

> launched into BP's objectives: independence, cash flow from the Forties Field, North Sea operations and international operations to be preserved. He said, 'The BP shares owned by the Bank of England are a problem. We can offer you help but no more.' When the question of independence came up, Harold got into a long, rambling metaphor about the virginity of BP, how the original marriage had not been consummated because the bride was frigid and how rape was involved . . . it was vulgar and thoroughly embarrassing. Steel was friendly but difficult. Pennell is a Tory industrialist and basically hostile. Frank seemed rather less than the full-sized man because BNOC is so weak and there was no indication of support from the PM.

Yet again, the inherent nature of power within the oil companies' gift, at virtually their permanent disposal, had been most effectively demonstrated. Nonetheless, this most combative of politicians, one whose abiding belief in the socialist ethic has never been in question, fought on, as his diary entry of 4 February 1977 makes clear. Once more, Tony Benn sought to extract from the oil majors the very best deal for Britain from North Sea oil.

> I had to deal with Amoco. On Monday night there had been long discussions with Amoco which went on till midnight and which were supposed to be concluded on Tuesday but had foundered on the simple point that the President of Amoco Europe, Mr Aune, the Amoco executive, Norman Rubash, and the lawyer, Ed Bissett, had declined to accept the form of words that Shell and Esso had accepted: namely that they would have a statement of intent as to their refinery policy in Britain and their readiness to try to optimise the use of North Sea oil; and that they would conduct their trade in a way that would maximise the benefit to the United Kingdom. They would not accept this because they claimed it would commit them legally to a £100 million investment in the Milford Haven refinery and they were not prepared to do that. So I said to Frank Kearton, 'Will you stick with me if I am strong?' Frank was terribly keen and the officials did, in effect, agree, so I called in Aune and Rubash at 11.15 without any of the others. They had been warned of the attitude I would adopt and I think

> they wanted to test it . . . I asked Aune and Rubash what had gone wrong. They produced a long explanation about how the whole thing had changed, they had never understood that what was wanted was this, and they had produced another draft. I said, 'Look, I am not negotiating it. We were going to settle the whole thing on Friday and we went through the words very carefully.'
>
> They said that the board of directors would never yield their powers over investment.
>
> So I said, 'You told me you were fully authorised to discuss it.'
>
> 'Well', they said, 'will you look at this new draft?'
>
> 'No. I cannot go beyond the Shell–Esso arrangements.'
>
> Rubash looked absolutely white. Aune looked shifty.
>
> I continued. 'This is it. You are dealing with Her Majesty's Government and these participation talks are intended to make a real difference. We are not prepared to be pushed. You are not dealing with a sheikh in the 1940s, you know, you are dealing with a British Government in the 1970s.'
>
> 'Well', they said, 'this hundred million clause . . .'
>
> 'I never mentioned a hundred million,' I said. 'You invented it and then you say it is a barrier. We have never asked that, but it is intended that there should be deep discussions about your market-policy . . . that is what the whole thing is about. You have to value the goodwill of the host government, and if you don't attach importance to that goodwill, that is a matter for you. Will you please let me know by tonight.'
>
> I was boiling with rage. I will never forget the experience with Amoco. If they won't cooperate, they won't get participation, and they won't get the licence, and that's it.

Such a cameo, illustrating the lengths to which the oil companies will go or, in this case, push their power, is of no small interest when attempting to understand the overall climate that prevails when the oil companies sit down with host governments to hammer out a deal, and is a particularly telling example of just where the real power lies. Of course, for those with fully focused eyes and fully tuned ears, and to those prepared to both see and hear, it has been long apparent as to who is on the podium conducting the orchestra when the men of oil play their tune, a state of affairs which prevails both at home and abroad.

CHAPTER 7

A POWER IN THE LAND

SURVEYING RURITANIA

A tide of sand flows towards us from distant dunes wrapped by the wind in winding sheets of emery-cloth. Rattling and shuddering over the bleached gravel of the desert road, the bus trails a drizzle of dust in our wake.

Here, on the lifeless edge of the Empty Quarter, words such as 'barren' and 'desolate' take on a new, sharper definition. Wendell Phillips, the oilman and historian who explored the region in the 1950s, called these pale grey flatlands between the dunes and the mountains 'the melancholy steppe of Oman', and the great desert traveller Wilfred Thesiger said this was a landscape that knew 'nothing of gentleness or ease'. Stepping down from a bus here is tantamount to being sandblasted in a sauna.

A shine of silvered steel in the middle distance provides a clue to what draws men to Lekhwair in north-west Oman. Flowlines, grouped and routed with finely calculated precision, turn through an acute angle to terminate in a remote manifold station, one of ten identical pieces of utilitarian engineering arranged in the sole-searing gravel with the abstract severity of modern sculpture.

For it is here, in this wind-raked, sun-blasted ocean of emptiness, where summer daytime temperatures reach 55° Celsius (131° Fahrenheit) that Shell, through its Petroleum Development Oman operating company, has constructed a $500 million temple of high technology in pursuit of oil.

Nowhere in the world has the getting of oil ever been easy. In Oman, it proved absurdly, almost perversely, difficult from the outset. And yet a study of the Sultanate reveals – and probably better than anywhere else on earth – how even the suspected presence of oil can transform a country, unseating its ruler and having him supplanted by his more malleable son, while turning the corporate finders and developers of the commodity into the real power in the land. Sultans may reign in Oman from glitzy palaces and adorn themselves with regal honorifics, but it is the sober suits from Shell who, in war and peace, have effectively set the agenda of government and, to a significant extent, dictated the nature and pace of development.

A British team began the search for oil with a survey in 1924–5, but with so many other and more immediately attractive prospects elsewhere in the Gulf, it was not until 1937 that a subsidiary of the old Iraq Petroleum Company (IPC) signed a prospecting rights agreement with Sultan Said bin Taimour. The concession area was later extended to include Dhofar in the south of the country.

The deal was, however, less than the sum of its parts. For the oilmen had stumbled into a mad, medieval Ruritania which had unaccountably, and by way of imperial India, the Zanzibar slave market and the staider shores of Whitehall, ended up with one foot in the Strait of Hormuz, the other in the Arabian Sea and both planted firmly in the unfathomable past. Nothing in Oman was ever quite what it seemed.

At the time of the first IPC deal, the Sultanate was recorded on maps as Muscat and Oman. The difference was more than a matter of nomenclature because it served to underscore how little of the ultra-secretive, Kansas-sized country was under the control of the Sultan, who was, technically, an absolute ruler.

In reality, the writ of Sultan Said, born in 1910 and educated at Mayo College in India, ran in only a third of Oman. Despite the lack of any obvious qualifications other than parentage, the young Sultan had, at the age of 22, been installed in Muscat in 1932 by the British, who had just jettisoned his father for financial incompetence. Britain, always anxious in the imperial past to protect its maritime trade and the sea route to India in particular, had entered into treaty arrangements and obligations with Oman centuries earlier. Whitehall noisily insisted that the Sultanate was a 'sovereign, independent nation', but everybody – especially in the Gulf – knew that in most important respects, the country was a British-subsidised colony in everything but name.

Even so, Said seemed, in the manner of Caligula's horse, an odd choice for the British to have settled on. An enthusiastic declaimer of Shakespearian verse, the new ruler spoke flawless English and Urdu but only uncertain Arabic in a country whose people understood little else. This was less of a handicap than might be imagined because Sultan Said made it plainly obvious on accession that he had no interest in, or intention of, engaging in dialogue with his subjects. Indeed, in his 38-year reign he addressed his people collectively just once, and then by way of a portentous document – 'The Word of Sultan Said bin Taimour, Sultan of Muscat and Oman' – nailed in January 1968 to the city gates of the capital. Since more than 95 per cent of the population were illiterate, this was a less than wholly successful exercise in mass communication.

Nevertheless, the top job in the Sultanate was anything but a sinecure even for a convinced and practising autocrat. Oman had also once been a colonial power with territory – Gwadur – in Pakistan and in East Africa too. It was Omani slavers who forced the Portuguese out of Zanzibar and Pemba in the sixteenth century and made the islands the market hub of their burgeoning business. Indeed, there had been a time when Sultans of Oman ruled their Arabian land from Zanzibar.

But all that was long gone when Sultan Said was presented with the throne and an almost empty treasury. For the truth was that Oman had been in a long and worsening decline ever since the abolition of slavery. One blow after another had fallen on a country now apparently ill-equipped by nature for anything more than reliance on subsistence agriculture and fishing; Oman might have had most of the world's finest frankincense but it had remarkably little of the world's potable water.

The descent into a new dark age continued throughout the war years until Oman, under the despotic, pathologically parsimonious, sometimes cruel, often petty and always eccentric rule of Sultan Said, became one of the most repressive and backward-looking places on earth. But it was here that Shell and some of the other major oil companies showed renewed interest in exploring for oil. In 1948, Richard Bird, an IPC representative, entered the country and succeeded in reaching Ibri in the Interior.

Bird, at the time, was travelling in a country that nationwide had fewer than ten kilometres of paved road, no electricity, no mains water, no radio, television or newspapers, no schools for those of its youngsters who managed to survive an infant mortality rate of 75 per cent, and just one tiny mission hospital in Muscat where expatriate Christian surgeons performed miracles by the light of kerosene lamps.

He would have seen extraordinarily few men wearing trousers, none in sunglasses, nobody smoking cigarettes and of the few able to read, none with a book tucked under an arm. The Sultan had at one time or another banned all of these pernicious influences. Like the loathsome Gómez in Venezuela, Said had also come to view education as a threat to his regime and a weapon to be denied to his people at all costs. He thought the London government should do the same. After all, he was fond of telling his British advisers, their insistence on educating Indians had cost them the Empire in the subcontinent.

Bird, however, had made progress of a kind and succeeded on his journey in securing safe passage for survey teams from each of the tribal sheikhs along the way. In Ibri, he even entertained hopes of being able to move on to Nizwa, the capital of the Interior, to conduct a face-to-face meeting with the Imam, Mohammed al-Khalili. The Interior, according to the terms of a treaty brokered by the British two decades earlier, was a self-governing theocracy in which the Imam combined leadership of Oman's dominant Ibadi Muslims with a large element of temporal control over the two-thirds of the country beyond the Sultan's power.

But the Imam, whose relations with the Sultan were often fragile, not only declined to meet Bird but also dismissed the whole idea of oil exploration as outrageously intrusive. Foreigners would not be tolerated in the Interior. Bird was sent packing.

The IPC man, it seemed, could do nothing right, for when Sultan Said heard – almost certainly from the Imam, the least welcome of all sources – about Bird's Ibri initiative, he flew into a rage. The oilman had exposed all too clearly the Sultan's inability to control events in his own bailiwick. Said bin Taimour's prestige had been dented and his ego badly bruised.

If nothing else, however, Bird's action stiffened the Sultan's resolve to unite, once and for all, Muscat and Oman under his rule. Though the Imam might at the moment be dismissive of exploration, just think what could happen were he to change his mind and the wretched Bird and his friends discover oil near Ibri or Nizwa. A stream – a flood, even – of revenues pouring into the Imamate's coffers beyond Sultanic control . . .

When the IPC man returned to Muscat six months later in a bid to placate the ruffled ruler, Sultan Said told him that he very much wanted oil exploration to go ahead but was under intense pressure from the Imam to keep non-Muslim Westerners out. And, as Bird had regrettably seen for himself, the Sultan lacked authority in certain key areas of the country.

Bird went away with the Sultan's written permission to re-enter Oman with another survey team in a matter of months. By and by, and this time accompanied by Said's personal representative, Bird returned but had barely set foot in the Interior when he and his team were fired on by the Imam's men and forced to beat a hasty retreat.

The stakes had, by this time, been elevated way beyond the matter of a minor turf war in a small, faraway place of which few had heard and about which even fewer cared. Shell and others were now serious about finding oil in Oman. And Said, assisted fortuitously by the Buraimi crisis which brought Oman and Saudi Arabia to within a whisker of war over the twin issues of oil and Saudi support for dissidents fiercely opposed to any extension of the Sultan's rule in the Interior, artfully manipulated the situation to ensure more tangible British help with unifying the nation.

The pace of events had been stepped up significantly. In 1954, Shell, Standard of New York, Mobil, Compagnie Française des Pétroles and BP formed a new outfit to go exploring for oil in the Sultanate. The company was called Petroleum Development (Oman) Ltd., the forerunner of PDO.

Two years later and 150 miles into the blast-furnace Interior at Fahud, the first PD(O) well – designated, unsurprisingly, No. 1 – was drilled. It was dry. Given the prodigious logistical and security difficulties that had been involved in the operation, most of the partners pulled out of the enterprise on the spot. Drilling the well had proved exceptionally expensive, and not least because a special security force – in reality, a private army directly and substantially funded by Shell – had to be employed to guard both the personnel and the masses of heavy equipment that had been landed on the coast of central Oman and, in a series of arduous and technically demanding operations, dragged ashore and hauled through the desert into the Interior.

Undeterred, Shell – driven as so often in its history by the hunger for crude – took over the partners' interests and went on with the search as 85 per cent owners of what had become the PDO company. By now, however, the security situation in the Interior had deteriorated into a real blood-and-bullets insurgency.

A MOUNTAIN OF TROUBLE

Ignited initially by the death of the old Imam and the election of the even more conservative and profoundly pro-Saudi Ghalib bin Ali al-Hinai as his replacement, the rebel campaigners – assisted by Ghalib's politically adroit and powerful brother Talib, governor of the nearby region of Rostaq – portrayed the presence of foreign oilmen in Fahud as evidence of an assault on the Interior's autonomy.

The insurgents made their military headquarters on Jebel Akhdar (Green Mountain), at 10,000 ft the highest peak in Oman and a formidable fortress. By now, the Imam's brother Talib had become the recognised leader of the insurgency and was styling himself Amir (Lord) of the Green Mountain.

With the Saudis looking on over one border and the first stirrings of what would ultimately prove a much more serious armed challenge in the far south, Sultan Said – aided by Shell's vigorous but discreet lobbying in London – now secured British military assistance on the ground and in the air. Towns and villages in the Interior from which the Oman Regiment of the Sultan's forces had been driven with such ignominy that the unit had subsequently to be disbanded were now bombed and rocketed by the RAF into submission.

Some idea of who was actually calling the shots by this stage of affairs can be gauged from a conversation that took place in Muscat between Sultan Said, his legendary military supremo Colonel (later Brigadier) Colin Maxwell and the 'man from Shell', Francis Hughes, the company's representative and general manager in Oman. The colonel put a question to his royal master about a particular aspect of the current military situation. The Sultan's reply was revealing: 'Oh, don't ask me. The man with the real answers to your question is Mr Hughes.'

The Sultan's – and Mr Hughes' – biggest military problem was that Talib's insurgents, having been blasted out of the towns and villages, were now ensconced high up on Jebel Akhdar where their discipline, tactics and staying power were impressing the British. Indeed, it was beginning to look as though they might hold out forever. Almost two years after the bulk of the Interior had been secured, it was clear that the rebel strongholds on the mountain would have to be overrun in a direct assault. Jebel Akhdar was successfully stormed, under cover of darkness, by the SAS on the night of 25–6 January 1959. It was a watershed operation for the regiment, one of whose troop commanders on the mountain was a young officer named Peter de la Billière. He was awarded the Military Cross for his part in leading the assault and, at the end of his career, would return to Oman after commanding British forces in the Gulf War as General Sir Peter de la Billière.

With the fall of the mountain fastness and the flight to Saudi Arabia of Talib to join his brother Ghalib, the insurgency was almost – but not quite – over. Small-scale skirmishing continued for several years and Shell-PDO oilmen and their contractors working deep in the Interior had to drive warily over roads still sometimes mined by rebels until the early 1960s.

Following the analysis of new survey data, the focus of exploration interest shifted in 1962 to an area called Yibal, 25 miles from Fahud. As at Fahud, the first well proved dry. The rig was moved a short distance and drilling began on Yibal No. 2. And almost 40 years after the search had begun, oil was at last struck on 18 September. The well produced clean oil at the rate of marginally more than 20 barrels per hour. It was expensive stuff: the long, difficult and, at times, dangerous quest had in total cost in excess of $20 million.

Events on a high, dessicated gravel plain in north central Oman might have marked the successful end of a particularly gruelling search, but they were only the beginning of the even more frustrating – and often bizarre – business of bringing an entirely new country into the international oil industry and turning the Sultanate into a revenue-earning exporter. The problems, even for a company of Shell's size, financial muscle and political clout, were immense. For Oman would have to be hauled out of medieval isolation, eased through a palace coup and a bitter guerrilla war, and somehow bounced into a semblance of mainstream twentieth-century life.

Following the initial discovery, more finds were made at Natih and, across a ridge barely a mile from the original dry hole, oil was located at Fahud in 1964. Work aimed at determining the extent of the oilfield pushed ahead and, by and by, the rig was driven into the stony valley where Fahud No. 1 was sited. And now, barely 150 metres from the well abandoned in exasperation by most of the original exploration partners, Fahud No. 18 proved both handsomely productive and an early object lesson in the maddeningly contrary and complex geology of Oman. Like the sea, the Sultanate would become notoriously reluctant to yield up its subterranean secrets.

THE EDGE OF ABUNDANCE

The prospect of oil coming out of the ground in commercial quantities focused attention on the absence of infrastructure. And because Oman had no roads, railways, airports – the British had fashioned an RAF base in Salalah, 1,100 kilometres south of Muscat, round a Second World War landing strip suitable only for military aircraft and there was another, even smaller, unsurfaced strip at Bait al-Falaj in the capital area – power stations or power supplies of any kind, construction of even the most basic facilities took on the air of heroic endeavour. There was not even a usable deepwater port; freighters of any size had to stand offshore and unload onto lighters.

Shell, as PDO, established a base camp of a few rudimentary cabins at Azaiba, not far from what is now the end of the runway at Seeb International Airport. But as the scale of operations grew, the oilmen sought a new and permanent base at Saih al-Malih, an exquisite bay – and favoured nesting ground of green turtles – giving onto deep water at the foot of the saw-toothed mountains surrounding the capital area.

It is unlikely that many of the people peering today through the gates of Al Alam (The Flag) Palace in Muscat realise that this singular building, dressed in a colour scheme of some eccentricity, resulted from a classic quid pro quo arrangement. Sultan Said was initially not at all happy with Shell's idea of turning Saih al-Malih into Mina al-Fahal, essentially an offshore anchorage with an onshore industrial area bristling with tank farms, workshops, a refinery, power station and office blocks. But when the oilmen suggested that they might in return for planning permission perhaps be allowed to fund construction of a modern replacement for the ramshackle and much patched Al Alam Palace, the Sultan graciously gave way.

Similarly, the man who regarded the spending of money as anathema did not struggle too hard against the creation by Shell of Oman's first thoroughly modern, fully equipped hospital. Ultimately, the Khoula facility, barely a mile from the heart of Mina al-Fahal, would be handed over to the Ministry of Health as a going concern of high reputation.

Later, under a Civil Assistance programme that Shell-PDO delivered with disarming generosity over three decades, many remote hamlets were connected to a network of graded roads with which the oilmen opened up the vast and hitherto inaccessible Interior of the country. The central and southern regions derived particular benefit from this development, for with the roads came access for scores of communities to power supplies and fresh water. To be sure, many of these schemes were small in scale, but their impact was disproportionately large among impoverished farmers and fishermen to whom the idea of life-enhancing development had previously been as insubstantial as a whispered rumour.

Work at Mina al-Fahal was geared to the arrival of a pipeline 156 miles long carrying oil from the Interior well sites to the tank farms at the coastal terminal. And here, in water 150 feet deep just 2,000 metres offshore, single buoy moorings (SBMs) were anchored to the seabed. Each of these 350-ton floating islands, fed by 40-inch pipelines from metering banks through which every drop of exported oil has to pass, was secured in position by 80 tons of chains. Supertankers, the colossal 250,000–300,000-ton vessels marine pedants insist must properly be called very large crude carriers (VLCCs), could then in operations of extraordinary delicacy be inched by pilots and loadmasters onto the buoys. When safely tethered, the leviathans could then be loaded with crude at the rate of 43,000 barrels per hour. Most of the vessels now load cargoes of between 500,000 and 750,000 barrels, the latter figure representing a day's current production in Oman.

When exports began, on 1 August 1967, the VLCC *Mosprince* steamed out of Mina al-Fahal laden with 500,000 barrels of Omani crude, a significant volume of oil given that production then was running at 160,000 bpd. And even that, in all of the circumstances, was no mean achievement.

For although Oman might at last have joined the oil exporters' club with its blend of medium-sour crude, the Sultanate remained the odd man out in every other way. The opening of an indigenous oil export industry had of course changed everything profoundly. Yet for the bulk of the Omani population, nothing was different. When men were walking on the surface of the moon, Muscat remained a gated, eerily quiet and dark city under permanent curfew. If you had been out beyond the city limits buying, perhaps, fodder for your donkey – still everyman's taxi cab and Transit van – and arrived back at the gate after nightfall, you were liable to be arrested and imprisoned.

The same fate befell anybody caught walking at night without carrying a lantern. And nobody made light of being incarcerated in one of Sultan Said's jails. Prisoners were always shackled and routinely tortured; they were only irregularly fed and watered. When after long years of grinding poverty oil was at last flowing and Oman stood on the threshold of abundance, traders in the main Muttrah souq were still using an antique system of weights and measures understood by few in the Sultanate and nobody outside it.

The matter of money was even more baffling. As oil revenues were piling up – they went from £1.4 million in 1967 to £22.5 million in 1968, £38.5 million in 1969 and £44.4 million in 1970 – currency and trade transactions remained subject to a startling array of imponderables and complications. In daily use within the memory of those now middle aged were Omani and Dhofari baizas, Maria Theresa Dollars (MTDs) divided into 120 baizas but actually traded at the rate of five rupees (which did not, of course, officially exist) to the MTD, while exchange rates in general were fixed in relation to the Kuwaiti dinar.

Every supertanker that steamed out of Mina al-Fahal left in its wake a country that still had only three schools and was governed – theoretically – by ministers, all of whom, with the single exception of the man holding the Interior portfolio, were British. In truth, many of the ministers had long gone and secretaries – or, indeed, *their* secretaries and assistants – were actually running things.

In a land still without newspapers, radio or television, life was apt to take on a truly surreal quality. Many perfectly sensible people came to believe that the Sultan had died years earlier, a fact hidden by the perfidious British so that they could rule in his name and trouser the oil money. Given that Said bin Taimour had neither been seen in public nor heard from in person since 1958, the notion was no more outrageous than the truth – that, discountenanced by the rebellion in the Interior, he had shut himself away in his palace in Salalah surrounded by a retinue of slaves and comforted by a harem of concubines: the Sultan told the writer Jan Morris that he liked 'a little loving in the afternoons'.

But in the south, many of the Sultan's subjects had lost patience with a reactionary ruler whose principal recreation, outside of the harem, was shooting at bottles arranged on the walls of his palace with an assault rifle. What would rapidly turn into a savage guerrilla war erupted in 1965 and, within four years, much of Dhofar was effectively under the control of revolutionary Marxist rebels who were able on occasions to mortar both Sultan Said's palace and the RAF base.

Unlike the Green Mountain rebellion, the insurrection in the south – begun originally in the name of the Dhofar Liberation Front but in 1967 taken over by the harder line People's Front for the Liberation of the Occupied Arabian Gulf (PFLOAG) – was not a muddled struggle in a faraway Ruritania. This was the brutally real deal in which well-armed, trained and ideologically motivated guerrillas – although entirely lacking air support – proved capable of taking on elements of the SAS, special forces sent by the Shah of Iran and King Hussein of Jordan, 'contract' soldiers from the UK and mercenaries recruited to the Sultan's forces from, among other countries, South Africa and what was then Rhodesia.

There was another important difference too in that Oman in the 1950s was a country of manifest, grinding poverty whose exports were limited to pitifully small amounts of fish, dates and limes. The Sultan was running the country on an annual budget – made up largely of revenues derived from an internal customs system of impenetrable complexity – which inclusive of British subsidies and grants most municipal authorities in middle England would have regarded as derisory. Military assistance during the Imamate rebellion had therefore been rendered free. But now, with substantial oil revenues in prospect and actually flowing from the summer of 1967, the Sultan was expected to pay his way militarily in the Dhofar war.

In truth, the old Sultan's days of power were numbered because his limpet-like adherence to 'old school' autocratic governance and extraordinary parsimony – he refused, point-blank, time and again to borrow against future abundant revenues money desperately needed for development – exasperated his few friends as much as it maddened his many enemies. Sultan Said had become a huge and indefensible political embarrassment to Britain and the focus of altogether too much hostile human-rights attention at the UN. With oil revenues between 1967 and mid-1970 totalling more than £85 million – and this in a country with a population estimated at barely 700,000 whose exports had never previously exceeded £1 million – Oman's backwardness could no longer be explained away, still less excused, by citing poverty.

On 12 June 1970, allies of PFLOAG attacked targets in Izki in the Interior close to the main pipeline and other oil installations. The raid concentrated minds powerfully and spurred Shell into lobbying in London for swift and effective action – both military and civilian – to put an end to the war. Eight days later, a general election in Britain put a Conservative government led by Edward Heath in power in place of Harold Wilson's Labour administration. With voting over, the pace of events quickened significantly. Stories began appearing in the British press – and, revealingly, in Iran too – suggesting that Said bin Taimour was about to be deposed.

GOODBYE TO ALL THAT

On 23 July 1970 there was a coup in Said bin Taimour's Salalah palace. Planned and perpetrated largely by British officials and officers, including the Sultan's Defence Secretary and the chief of his intelligence service, there were times when this long drawn-out affair, punctuated periodically by the sound of small-arms fire and fleeing footsteps in corridors, seemed to have been stage-managed by the Keystone Cops with a little help from Gilbert and Sullivan.

Always described subsequently by the Oman government's expatriate spinners as the 'bloodless' or 'nearly bloodless' overthrow of father by son, it was nothing of the sort. There was an immediate casualty when one of the Sultan's servants was shot dead in the initial confused gunfire. A little later, Sheikh Braik bin Hamood bin Hamid al-Ghafri, the courageous son of the governor of Dhofar who had been deputed by the plotters to face the Sultan and demand his abdication, was shot in the stomach in the course of further confusion. The Sultan was himself armed and fired his revolver four times. It is somehow hugely and sadly symbolic of what turned into an Ortonesque black comedy that with the fourth discharge of his gun, Sultan Said shot himself in the foot.

With the eventual departure of Said bin Taimour by plane for London and discreet exile in the sybaritic sanctuary of the Dorchester hotel, his Sandhurst-educated son Qaboos bin Said was thrust onto the throne.

The new Sultan, having inherited a treasury bulging at the seams with oil revenues, immediately set about an immense and wide-ranging programme of reforms aimed at nothing less than the transformation of Oman. Schools, hospitals, clinics, roads, ports, airports, power stations, radio and television services and a government-controlled press were all begun in a frantic burst of activity. There were indeed times when Muscat came to resemble Pithole and Spindletop as vast amounts of oil-derived money – the government was taking 50 per cent of all PDO-generated profits plus 12.5 per cent of all crude exported – coursed through the economy and out of Oman, without, it sometimes seemed, touching the sides. By 1972, imports had rocketed from £4 million in 1966 to £55 million; the number of motor vehicles being brought into the Sultanate had increased by a dizzy 600 per cent. 'In such conditions,' said the *Financial Times*, 'it would be difficult for anyone with a modicum of common sense and experience of the market to fail.'

Marxist analysts saw things somewhat differently and one, having commented on the number and well-connected nature of the snouts being buried in a mighty trough, observed that 'Oman's oil income is available for all the world to pillage.' And it was undeniably true that Britain had reaped a magnificent reward from the coup by grabbing within two years a whopping 27 per cent of the Sultanate's entire import market.

Qaboos's coup was formally welcomed and saluted by Shell in a statement released by PDO's managing director on 1 August 1970, exactly three years after oil exports had begun. And it was the new Sultan's great good fortune that oil production and revenues were able to keep pace with his government's incendiary spending – or at least make Oman a better and more creditworthy risk in the long-term loan markets to which Qaboos would have frequent recourse.

Gradually, bloodshed in the south yielded to a skilful 'hearts and minds' campaign and Dhofar was pacified. Oman was at last united under a Sultan who, if not universally popular – some believed there were better claimants to the throne, there were others who yearned for a theocracy and still others who simply hoped for a socialist end to Sultanic rule altogether – was recognised in the north and south, in the Interior and on the coast. Even the disgruntled had the consolation of knowing that the new incumbent was indubitably alive. Qaboos, however, soon managed to upset both traditionalists and his more radically inclined subjects by arrogating to himself the prefix 'His Majesty'. None of his 13 Al Said dynasty predecessors had ever aspired to anything grander than 'His Highness'.

It quickly emerged that peace had been achieved in Oman at a fearsome cost. Spending on defence and the paramilitary police – who in the Sultanate have armoured personnel carriers equipped with heavy machine guns and water cannon – from the outset swallowed more than half the government's annual expenditure.

And soon – very soon – after the coup, it looked as though the revenue-earning potential of the country's oil industry might not be as secure as had been forecast. For production actually fell from 121.3 million barrels in 1970 to 107.9 million in 1971 and 103.2 million in 1972 before beginning to rise again. In truth, some pundits were already starting to write Oman's oil epitaph within five years of the industry's start-up.

The reason for the sharp decline in production was either commercial acumen or naked greed, depending on your viewpoint. Production had originally been scheduled to begin at the end of 1967 but, in June of that year, following the Arab–Israeli war, Arab oil producers imposed a temporary boycott. There was thus a gap in the market that Shell-PDO sought to partially fill with Omani medium-sour crude.

Also, as though to underscore the intensely political nature of the oil business, the Biafran War had seen the disruption of exports from Nigeria, which again gave Oman a newcomer's toehold in the market.

Exploiting these opportunities had, however, extracted a familiar price in that well pressures were seriously disturbed by over-hasty production: too much oil was pumped too soon. All over the world, the industry's history has been shot through with instances of perceived short-term gain causing long-term damage to even the most prolific oilfields.

And now, as exploration activity was again increased and new fields urgently sought, PDO – well on its way to its current status as a company of which the Oman government owns 60 per cent, Shell 34 per cent, Total 4 per cent and Partex 2 per cent – began to discover and confirm some uncomfortable truths.

For Oman, so different culturally from its neighbours and everybody else on the Arabian peninsula, was also different geologically. According to the authoritative *Alexander's Gas & Oil Connections*, Omani wells, on average, produce only one-tenth of the volume of wells in neighbouring countries. Moreover, Omani oilfields are smaller, more widely scattered, less productive and much more costly per barrel than those of other Gulf countries.

RECALCITRANT RESERVOIRS

All of this made Oman the test-bed for what Shell calls 'enhanced oil recovery' (EOR) technologies and techniques. These are measures aimed specifically at wringing the best possible performance from recalcitrant reservoirs as well as lifting the production of fields moving towards exhaustion.

Besides horizontal drilling – a primary EOR ploy – PDO has injected and flooded some of the Sultanate's oilfields with water, pumped polymer under pressure into others, soaked Amal's heavy crude-bearing formations with steam and even resorted to the electromagnetic heating of reservoirs in a bid to improve the flow and recovery rate of oil.

It seems paradoxical that in attempts to reduce already high production costs, vastly more money must be spent. For EOR engineering is fiercely expensive. Lekhwair was the place where Shell elected to deploy the entire box of tricks. A smallish field producing about 24,000 bpd of (highly corrosive) hydrogen sulphide-soured crude, Lekhwair's oil is locked in a tight limestone formation.

To get at it more effectively and make best use of the abundant gas associated with the oil, Shell decided to fracture the reservoir by the injection under high pressure of massive amounts of water. This would sweep the oil towards a total of more than 170 wells – some of them horizontal, almost all of them newly drilled – with the aim of nothing less than the quadrupling of production to 100,000+ bpd. With the water taken into account, a total of 160,000 bpd of liquids would be coming out of the desert ground.

It sounds beguilingly simple. The reality was dauntingly complex and its achievement involved a hatful of industry 'firsts'. For never before had high-pressure reservoir fracturing been attempted in a tight limestone formation, nor had the technique been employed on anything like the scale of the Lekhwair project.

Similarly pioneering – if more mundane-sounding – was the use of 16-inch plastic pipes to feed crude oil into the production station, and polyethylene sleeving in the water injection ring to prevent corrosive products entering the reservoir from the injection wells.

But beyond a bunch of plastic pipes and masses of fractured rock you can't even see, what do you get for half a billion dollars in a burning gravel desert? The answer is a sort of heavy engineering Disneyland with

400-ton mobile drilling rigs – designed in France, built in America and capable of being hauled with their 50-metre-high masts erect by monster prime movers between well sites – and microelectronics of fiendish cleverness and sophistication in a touch-screen computer control system.

Some idea of the scale of the project can be gained from the litany of Lekhwair numbers: 175 specialist design engineers and 1,650 construction men spent eight million man hours building the plant and facilities. Contractors and PDO staff involved in the work drove between them a total of seven million kilometres. Fifteen thousand cubic metres of concrete were poured into the project and 50,000 tons of equipment brought to the site. Getting the materials into Oman required 300 ships and 200 aircraft.

Muscle for the production station where oil, gas and water are separated and processed is provided by four gigantic Borsig compressors. Of 10 MW each, these huge machines are driven by 70-ton electric motors via hydraulic couplings similar to an automatic transmission in a saloon car except that they are almost exactly 1,000 times bigger.

It goes almost without saying that Lekhwair has its own 50 MW power station – big enough to meet the energy needs of a European industrial town – and that those employed at what has gone into the record books as the hottest major worksite on earth are housed in a 170-man permanent camp with full (and essential) air conditioning, a swimming pool, library and neat shrubbery kept green and growing by recycled waste water. The camp is an oasis of no small comfort in this most unrelentingly hostile of desert environments.

That Lekhwair, together with other EOR projects elsewhere in the Sultanate, has worked is evidenced by *Alexander's* judgement that production costs have been reduced to $4 a barrel in some fields and $3 in others. These figures, though good by world standards, are still substantially higher than those of other Persian Gulf oilfields. PDO has recently stated that it aims to double its recovery rate to 50 per cent by investing another $1 billion over five years in a Target 50 programme. *Alexander's*, commenting on this bold pronouncement, says flatly: 'It remains to be seen if this will be possible.'

If Lekhwair has exemplified Shell's willingness to invest heavily in high technology in pursit of increased production in Oman, Yibal – as the reserves crisis has made cruelly clear – serves now to symbolise the Sultanate's unique capacity for deflating the dreams of ambitious oilmen. For having achieved the status of the country's most prolific oilfield producing, at its 1997 peak, 220,000–250,000 bpd, Yibal is now Shell's biggest problem in the Sultanate.

Production from the celebrated field is in a tailspin and has declined at an annual rate of 12 per cent – more than twice the regional norm – to its current level of 88,000–90,000 bpd. And Shell has also been obliged to admit the existence of another, closely related, major problem in that 90 per cent of what is now coming out of the ground in Yibal is water. This is adding significantly to Oman's already high – by Gulf standards – costs of production. Yibal's decline has signalled to industry analysts the far more worrying prospect of the failure of Shell's much-trumpeted horizontal drilling strategy to deliver the promised and heavily hyped benefits.

In essence, and as its name suggests, horizontal drilling is a technique for turning a well – usually over a radius of 150 metres – through 90 degrees to run parallel with the horizon for anything up to a kilometre at whatever depth is required.

Compared to a conventional, vertically aligned well, the horizontal section permits the drainage of a much greater area. The efficiency gains are often spectacular and two- and even three-fold increases in production are commonplace. Shell did much pioneering work with horizontal drilling and made big claims about the capacity of the technique to prolong the life of 'mature' fields and boost productivity in others.

But as in life, so in oil: there is rarely gain without pain and horizontal drilling is technically challenging and, even by oil industry standards, expensive. You need to be neither a technologist nor an anorak to appreciate that trying to control a diamond-tipped drill bit describing a giant capital L through rock a mile and more underground is not for the faint-hearted or those operating on limited budgets.

The technique has had no more enthusiastic champion than sacked Shell chairman Sir Philip Watts. In a breezy, generally upbeat public report in 2000, he identified it as the key element in 'major advances in drilling' that were enabling the company 'to extract more from such mature fields [such as Yibal]'.

When Watts made his remarks – as Shell's head of exploration and production – he must surely have known that production at Yibal had plunged by almost a quarter in two years. And he should also have known a year earlier – when he was cheerfully talking about 'advances in well technology' that were producing 'substantial additional reserves' – that horizontal drilling, far from being the solution to Yibal's decline was, in truth, a significant part of the problem.

The evidence had been provided by two Omani petroleum engineers who, in reports published by PDO internally and as technical papers by the Society of Petroleum Engineers, highlighted the fall in production at Yibal and the 90 per cent water problem. One of the papers went on to spell out the fact that this was a consequence of the volume of water being injected into the field as an integral part of the horizontal drilling programme. This was not welcome news, since PDO had drilled almost 100 horizontal wells at Yibal in a $200 million development project in 1996.

If Watts had wanted another – heavyweight – technical opinion about what was happening deep beneath the gravel desert, he had only to turn to Walter van de Vijver, the man who became Shell's head of exploration when Watts was himself elevated to the chairmanship. For van de Vijver, widely believed by Shell insiders and informed City specialists to be Watts' anointed successor, knew all about Yibal, having been in charge of production in PDO's northern oilfields earlier in his high-flying career.

It is now clear from their barbed email exchanges that the two men spoke to each other as little as possible, so Watts might instead have preferred to seek the views of Malcolm Brinded, the man who took over as chief of exploration when van de Vijver was forced out in Watts' wake.

The Cambridge-educated Brinded, a highly regarded oil industry professional who had served on secondment as an adviser to the government's Department of Energy, was earlier in his career PDO's director of development with responsibility for, among other fields, Yibal and Fahud. Brinded had also served as PDO's chief of engineering and had headed some huge high-technology oilfield projects, including Lekhwair.

And if Watts had not wanted to go down the line for the opinions of men with hands-on experience, he could have gone upwards to seek the view of his illustrious predecessor, Sir Mark Moody-Stuart. For he, too, was a PDO veteran and, moreover, the author of technical papers on Oman's oilfields.

It was of course no coincidence that so many of Shell's current top management – and the men who had been their bosses, such as Sir John

Jennings, the company's tenth chairman who in 1962 and at the age of 25 had been the well-site geologist at Yibal when oil was at last found – had spent important parts of their careers with PDO. A management training consultant much used by Shell said: 'I think the company recognised very early on in its relationship with Oman that the country was actually a giant adventure playground for bright young managers. Given the extraordinary mix of problems – both technical and what you might call ethno-political given that relations between the tribes in the Interior and coast have not always been harmonious – there was a general sense that if you could hack it in Oman, you could hack it anywhere.'

But even as Shell tied itself in a technological knot and seriously overstated Oman's reserves, as well as making a maladroit nonsense of its handling of the whole affair, there was no doubt that the company and Oman were prospering greatly from PDO's major natural gas finds – often deep beneath old oilfields – and the development of LNG exports.

The first big discoveries of gas were made at a depth of 4.5 kilometres in Barik and Saih Rawl in central Oman. These, plus a smaller find in Saih Nihayda, formed the basis of LNG exports in a project developed by the Oman LNG Company, a joint venture 51 per cent owned by the Oman government, 30 per cent by Shell, 5.5 per cent by Total with the rest of the stock being owned by Partex, Korea LNG, Mitsubishi, Mitsui and Itochu.

The liquefaction plant from which exports began in 2000 is located at Qalhat, near Sur, on the Sultanate's east coast. Both were port towns of importance in the slave- and spice-trading days of Oman's former and long-lost prosperity.

A RACE OF GLOBAL GIANTS

Some 75 years ago, A.C. Hardy, the European director of World Petroleum, wrote: 'Oil belongs to a democratic age and has been developed along essentially democratic lines, although in doing so it has tended to create an entirely new aristocracy.'

A little earlier, R.G. Hawtrey, an economics chief at the British Treasury, wrote in his *Economic Aspects of Sovereignty*:

> The profit-seekers are usually in a position to exercise influence over their own governments, and governments regard the support of profit-seekers' activities in every part of the world as a highly important aim of public policy . . . The gain of one

> country is necessarily loss to others; its loss is gain to them. Conflict is the essence of the pursuit of power.

Both Hardy and Hawtrey are long gone, but their analyses remain relevant and valid today. For when words composed to support self-interest, and the cant and high pretence that attend it, are swept away, what is left is the raw power exercised by Hardy's 'new aristocracy' and Hawtrey's 'profit-seekers', who in the extraordinary world of big oil can, and do, order the economies of states as well as dominating their political agendas.

In his 1975 book *The Seven Sisters*, the distinguished British writer and oil industry historian Anthony Sampson commented:

> The band of sisters has been led by two giants, Shell and Exxon (formerly Standard Oil) . . . Their rivalry across continents has long been a sub-plot to modern history, financing whole nations, fuelling wars, developing deserts. Their commercial ambitions are fraught with diplomatic consequences: the revolutions in Iraq, the separatist movement in Scotland and the civil war in Nigeria. They had seemed often enough like private governments, to which Western nations had deliberately abdicated part of their diplomacy . . . they represented much more than themselves; they were a central part of the whole economic system of the West.
>
> Their incomes were greater than those of most countries where they operated, their fleets of tankers had more tonnage than any navy, they owned and administered whole cities in the desert. In dealing with oil, they were virtually self-sufficient, invulnerable to the laws of supply and demand and to the vagaries of the stock markets, controlling all the functions of their business and selling oil from one subsidiary to another. Shell oil was pumped from Shell oilfields into Shell tankers, on to Shell refineries, and through Shell pipelines to Shell filling stations. They were the first global giants.

Sampson's theme was taken up later by Said Aburish, the celebrated writer on Arab affairs, in his work *The Rise, Corruption and Coming Fall of the House of Saud*:

> Acting in concert with, and backed by, their governments, they [the oil majors] have as their sole aim to maximise their profits against the interests of oil-producing countries. Throughout the Middle East, they have arrogated to themselves the following rights: where to explore for oil, how much to invest in oil exploration, how much oil to produce, what price to charge, how to share the proceeds among themselves, how to transport the oil and what political leaders of the oil-producing countries to support or oppose.
>
> Given that oil is the primary source of livelihood for most oil-producing countries, the extent of influence, indeed control, the oil majors exercise over the destinies of the producing countries was a unique historical situation, the like of which the world is never likely to see again. Indeed, given that in the 1940s and the 1950s the oil majors controlled Iraq, Iran, Venezuela, Saudi Arabia, Kuwait, Nigeria, Indonesia and other countries, we are entitled to speak of an oil empire.

The question, in view of the track record of oil companies like Shell behaving as sovereign states within sovereign states, is how they continued to manage the trick. The answer, in large measure, is brutally simple: size. By the end of the First World War in 1918, Shell was operating with an annual budget well in excess of that of some European states. By the end of the Second World War – and, as Anthony Sampson noted, with a tanker tonnage double that of most navies – Shell's budget would dwarf those of most of the smaller member states of the UN.

When Shell or any of the other majors walked into a country such as Oman, they got what they wanted, when they wanted it, because they had the resource muscle to make or break governments, mobilise private armies, start or stop wars and insurgencies, educate the masses or simply make a few fat dictators even richer and more corrupt. It was never a matter of playing by the rules because the oilmen wrote the rulebook wherever they went.

To adapt the words of the Great Helmsman Mao Tse-tung: power comes from a barrel of oil and a Shell lawyer's pen. Anybody who is still unsure about who wields the biggest stick in many lands and economies should consider Shell's own words on the matter. For in the section dealing with 'Arrangements with Governments', the company's invaluable *Petroleum Handbook* says under the subheading 'Fundamental Aspects':

> The management or the control over operations may range from virtual freedom under a concessionary or lease arrangement, without state participation, to little or no control . . . Very generally, the degree of control left to the oil company will be commensurate with the degree of investment risk undertaken . . . Under a concession or lease the producer will obtain the totality of production but may have to offer a proportion for sale to the NOC [national oil company]. In the case of state participation such a right will be proportionate to its equity share in the joint venture . . . Clearly, while the economic return aspect is the decisive factor in reaching an exploration and production agreement . . . the economic results of the venture as a whole will vary considerably from case to case; as a consequence, the 'government take', that is the share of the economic results accruing to the host state by way of such items as taxes, royalties, profit sharing and production sharing, will also vary.

So far, unexceptionable. But under the no-nonsense heading of 'Other Factors', the *Handbook* gets to the heart of the matter:

> To afford protection to the investor, there will be freedom from certain duties and taxes; freedom to remit and dispose of profits; security of tenure; and provision for applicable law and international arbitration . . . In many older production-sharing contracts, the oil company, while liable for income tax, will not actually pay the tax; it is included in the government's profit share . . . taxes are alleviated by the introduction of measures reducing taxable income, for example, by excluding part of the proceeds from the calculation, or by allowing additional deductions expressed as a percentage of deductible expenditures, resulting in a deduction of such expenditure more than once.

It is, as Shell's colossal profits have over the years demonstrated, nice work when you can get it.

CHAPTER 8

SPIN, SMOKE AND MIRRORS

The man who would be Poet Laureate was fulsome indeed in his praise, unstinting in his gratitude to those who had made the whole enterprise possible. The year was 1934, the time of fascist dictators had arrived, the people of the European continent were sleepwalking to war, but in Britain, the age of innocence, if on occasions contrived, continued to reign supreme. The King was on the throne, the Royal Henley Regatta had brought small boats and yachts on to the placid Thames, there was Royal Ascot to look forward to and summer holidays to be enjoyed at the coast. The sun shone overhead and all was, for the while at least, well with the world. And in the glorious June of that same year, Shell had made its very own contribution to such a state of well-being by financing the writing and publishing of a *Shell Guide to Cornwall*, that 'other world' where south-western England finally runs out to the vagaries of the wide, rolling Atlantic Ocean.

The editor of this latest public relations exercise by Shell was the poet John Betjeman, who, some 38 years later, was to be appointed Poet Laureate by Queen Elizabeth II. Betjeman was well known for his abiding love of the English countryside, its fields, hedgerows and forests; and principal in his affection was Cornwall, where he lived during the autumn of his life and where he now lies buried in its ancient soil. He was a sensitive soul, kindly and with an air of gentleness. As the years progressed, he was to recoil in undisguised dismay and horror at the vulgarity and brashness which, today, so assails all he loved in his native land. He was also well associated with generosity, even when writing poetry was not bringing in the pennies in adequate amount, and both aspects of his personality appeared in his preface to *Cornwall*.

> I must thank the enterprise of Shell for sponsoring such a book. By so doing they are, it is hoped, furthering the interest in English scenery which their many excellent posters have encouraged and acting with the disinterestedness which caused them to take down disfiguring tin signs some years ago and to set an example to other companies.

And just in case the readers of Shell's *Cornwall* had missed John Betjeman's message on page one of the guide, the last page served to make the point yet again. In a pen-and-ink drawing by one of Shell's superstar artists, Edward Bawden, two Cornish fishermen appear, suitably attired, one dancing his way along a quayside holding a concertina above his head, the other sitting on a lobster pot playing a mouth organ, while, in an outburst of environmental licence and excess, two mermaids look up adoringly at the melody-making pair from the waters of the harbour.

But it is the words that constitute the caption to Bawden's highly improbable Cornish cameo (even in the halcyon days of 1934) which seek to activate the taste buds of the readers' minds, to impregnate their subconscious with the message that you can, indeed, always place total reliance on Shell, once, that is, you have put its very fine product in the petrol tank of your very own car: 'Lanteglos-By-Fowey but Motorists buy Shell. You can be sure of Shell.' It is not the clever play on words – 'By' and 'Buy' – which is of particular interest but the slogan, which was, in 1934, in its infancy as a marketing strategy. It was a neat arrangement of a few words that, across the generations, proved to be a most successful public relations ploy, if not one of the most successful ever produced by the world of commerce: 'You can be sure of Shell.' Indeed, its great success is best understood when the fact is recorded that in today's fast and furious times, cars on the garage forecourts of the world have their tanks filled with Shell at the rate of one every four seconds of every day, while civil airliners are similarly filled with Shell aviation spirit every ten seconds.

It is also of special interest that in the preface to the guide, Betjeman makes reference to Shell '... furthering the interest in English scenery which their many excellent posters have encouraged ...', although his follow-on statement on the company's 'disinterestedness' in being fired by the promise of increased profits as a result of their apparent concern for the English landscape and its fauna and flora can but be regarded as a sign of this most gentle man's innocence and, perhaps, a wholly understandable naivety regarding the ways and wiles of the corporate creed of enlightened self-interest. The fact remains, however, that the slogan did indeed inspire a series of posters and billboard advertising campaigns which are seen today as works of art in their own right. On the rare occasions that Shell posters from the 1930s come onto the open market, they now command very high prices indeed.

The slogan had various methods of presentation in addition to the posters, with, almost certainly, the most celebrated being the American crooner Bing Crosby singing the words in his usual 'smooth as silk' style and internationally recognisable, mellifluous tones: 'Keep going well, keep going Shell, you can be sure of Shell, Shell, Shell.' In a word, or two, it worked. The world that was to come to regard petrol as the key to the realisation of the age-old dream of personal, private mobility (and, of course, so much more besides) was lulled into the 'soft' land of Shell: the pastel-coloured posters that extolled the beauty and virtues of England's countryside, a barn owl looking down from a hollow in a glorious towering oak at a happy family motoring through the verdant landscape below. The guidebooks, such as Betjeman's *Cornwall*, included avuncular advice for those planning a picnic: 'Don't forget a rug to sit upon, the salt and sugar, to wrap up your sandwiches in greaseproof paper ... oil of lavender and citron to keep the flies away and a blue-bag in case you get stung by a wasp.' And then, as if an afterthought, '... a spare can of petrol'. Now that was the real 'sting' in the tail of Shell's most famous slogan, one that the international family of humankind was not expected ever to forget.

The company was also a pioneer in the field of sports sponsorship, eagerly acknowledging the universal truth that if, indeed, the way to a man's heart is through his stomach, then his heart can be guaranteed to beat just that little bit faster at the prospect of being associated in virtually any manner with those who excel in any form of sport. This being so, it was a 'natural' for Shell that the company should take up financial sponsorship of such titans of the motor-racing track as Stirling Moss. Indeed, and well before the contemporary world of corporate madness and disfigurement had descended upon an increasingly hapless world, where all sorts of sponsorship takes place – from ego-crazed individuals who attempt to reach the North Pole on a pogo stick to the planting of company billboards on roundabout 'gardens' – the yellow-hued shell of Shell, over which, not so incidentally, the company holds worldwide copyright, became a prominent feature of the motor-racing track, with Stirling Moss regularly assuring the world that he wouldn't dream of leaving home, or the pits, without a car that had been liberally lubricated with Shell X-100 Motor Oil.

As a marketing strategy it worked not just well but very well indeed. It was, in actual fact, and in tandem with the success of the company's guidebooks to the countryside, a strategy of brilliance, one that in a most effective style and manner encapsulated and mirrored the world according to Shell: a green, clean, comfortable and safe place whose only variance was, in the case of motor-racing sponsorship, that it permitted those who were wise enough to purchase the company's products an association with the exciting and very sexy world of high-speed sport. It remains a principal feature of Shell's marketing strategy, with the company's sponsorship of Formula 1 drivers such as the Stirling Moss of the contemporary world of the motor-racing circuit, Michael Schumacher. It is an advertising strategy of brilliance principally in that, by tapping into an aspect of human activity the appeal of which is both ageless and universal, establishing human prowess via physical skill, stamina and endurance, its message can be carried down and through the generations: you can, well, be sure of Shell however fierce the competition and whatever the weather.

The poster and billboard campaigns, allied to the company's related sponsorship and publication of a whole range of books designed to encourage the motorist to get out more, worked a very special kind of magic with the petrol-consuming public. The message was clear and there for all to see.

The 'father' of such deft and appealing campaigns was one Jack Beddington, whose inspired and highly successful reign as Shell's advertising manager ran from 1932 to 1939. According to contemporary accounts, he was as far from the purple-shirted, glib-gabbling, nouvelle-cuisine addicted, second-home in Tuscany, machine-tanned hullabaloo

PR practitioner of today as it would be possible to find. He was, ran a contemporary comment:

> a man who gave all the appearances of a Whitehall mandarin, a Treasury solicitor perhaps, with all the airs that one would expect from that class of Briton in the 1930s. He was quiet to the point of virtual invisibility, with the clipped vowels and somewhat stiff attitudes of an individual living in the Home Counties, attending to his prize Lobelias at the weekend and, on occasions and when suitably fired, writing letters to *The Times*, which he would sign as 'I am, Sir, Your Most Obedient Servant'.

But if appearances are, indeed, deceptive, then Jack Beddington was a particular case in point. Mr Beddington was by no manner of means anyone's 'Obedient Servant'; he was, according to Shell legend, very perturbed. The source of his disquiet was the nature of Shell's advertising material, and far from being the retiring individual so many thought him to be at first acquaintance, he marched, uninvited and unannounced, into the company's London headquarters and told them so.

In terms both clear and unequivocal, Jack Beddington told the men at Shell that their advertisements were all too often long-winded, tedious and larded with ineffectual, scientific jargon which left the average member of the public not just cold but alienated from the product he was meant to purchase by dint of the advertisement's appeal. The 'science' of selling petrol to consumers all too ready to buy it, declared a somewhat confident Beddington, was not, in actual fact, rocket science. It had to be kept plain and simple, couched in an arrangement of easy-to-read words making it easy to understand and, most crucially of all, to remember and relate to. And, continued this genius of the advertising age, it was not simply down to a question of a few carefully chosen words. Pictures, pretty pictures, would have to have an awful lot to do with it as well. The modern motorist, concluded this evangelist of good public relations, was little interested in the molecular structure of the petrol they put into the tanks of their cars. What they needed to motivate them to become legions of loyal Shell consumers was not scientific facts but, as they embarked upon a Sunday afternoon drive in the family saloon car, with the open road ahead to the broad, sunlit uplands of the English countryside, an assurance that they could indeed be sure of Shell.

Whilst no record exists as to how long it took for the men at Shell to realise that a genius in the making, a new prophet of the power of advertising, had stood before them, it is known that they wasted little time in offering him the post as the company's advertising manager. What is also well established is that Jack Beddington, armed with a budget that was the envy of others in the advertising profession, lost no time at all in recruiting to the service of the yellow Shell artists of international calibre such as Paul Nash, Ben Nicholson, McKnight Kauffer and, of course, that master of the pen-and-ink genre, he of the squeeze-box playing, dancing Cornish fisherman, Edward Bawden. The high fees they commanded, and which Beddington was only too pleased to pay, were, in double-quick time, seen to be well justified, as Shell's poster and billboard campaign became as common a feature of the English landscape as the steeple on the village church, the swinging pub sign, the red pillar box, the bobby-on-the-beat and the market square. The campaign was a brilliant success, just as Beddington had known all along it would be. It was a triumph and set a standard in national advertising that was to last for half a century and more.

The men on the company bridge at the state of Shell were, not unnaturally, pleased beyond measure at their 'find' in Jack Beddington, and not just because his brilliant advertising campaigns harvested ever-growing profits. And here lies the essential factor, some would say not so much the missing factor in such an equation of unparalleled success but one that, upon even the most cursory examination, is all too obvious. Shell actually needed little advertising exposure to ensure successful sales. Indeed so vital a commodity had petrol become, not just to industry but also to the quality of domestic life as epitomised by the family car, that little energy was really required to 'push' the product in the marketplace. There never was the need for Shell to have to try too hard to sell its products, beyond, of course, establishing a nationwide network of forecourts into which a never-ending stream of vehicles would arrive in order to fill up their petrol tanks. No, the essential service provided by Jack Beddington's genius was that it gave the industry an acceptable public face. Shell, via the manipulation of public acceptability, achieved universal approval through a poster campaign that suggested that Shell products were 'green' and they were 'clean'.

And, given the mighty profits which have endured for almost a century, the company has always been only too ready to assume the additional role of a publisher of often lavishly illustrated books extolling the beauty and diversity of Mother Earth, from the lions and leaping antelopes of the African high savannah to the bird-life of Britain, and much else besides, as long, that is, as such publications could weave into public consciousness the message that Shell entertains a commitment to the natural world, to which its many and varied products do not constitute an environmental threat.

Its financial backing of such publications has been, as with the company's celebrated poster and billboard campaigns, an outstanding success, with a high degree of quality of style and presentation being a constant feature, a fact best illustrated by recording the fact that, in a continuation of what can be justly termed the 'Beddington factor', only writers and artists of the highest talent, such as Geoffrey Grigson and David Gentleman, have been recruited.

THE DREAM MACHINE

What many could not see, however, was that such advertising campaigns concealed the true state of affairs; that, in actual fact, all was far from well in the rolling forests and green pastures so exquisitely portrayed by Shell. Indeed, the company billboards most effectively shielded from public gaze what was a truly grotesque picture. The carefully (and cleverly) crafted image was a sham and a particularly disgraceful one at that.

Take just one example. At the very same time the company was publishing some of the finest and, indeed, strikingly beautiful, evocations of the English countryside in its 'Everywhere You Go' campaign, an aspect of which was a portrayal of some of England's finest birds of prey – the heart-stopping beauty of a barn owl in close-up and a kestrel on the wing – it was producing, and selling hard to the country's agricultural establishment, some of the most toxic and persistent pesticides known to man and beast, such as Aldrin, Dieldrin and Eldrin. Indeed, so intense was the toxic nature of these pesticides that they dealt an almost terminal death blow to many species in Britain's predatory bird population. It was a catastrophic state of affairs and, yet again, a clear contradiction of everything Shell sought so hard to have its customers believe.

As well as having a devastating impact on the British wildlife, the effects were also being felt in America as much of the experimental stage of development of the company's Drins took place at a secret location in the state of Colorado, Rocky Mountain, near Denver City, a site previously used by America's military establishment to test deadly nerve gas.

Shell's Rocky Mountain experiments on its Drins did not take long to produce a truly deadly and quite dreadful effect on the local wildlife and the whistle of warning was blown by a game warden of the Colorado Department of Fish and Game who had carefully documented abnormal behaviour in the local wildlife. He voiced his concerns to Shell personnel working at the Rocky Mountain site but his appeal met with no sympathy whatsoever, and his anxieties were dismissed as being of little concern.

Writing in the London *Observer* newspaper in 1993, the celebrated journalist Adam Raphael commented:

> By 1956 Shell knew it had a major problem on its hands. It was the company's policy to collect duck and animal carcases in order to hide them before scheduled visits by inspectors from the Colorado Department of Fish and Game.

Such a sorry episode assumes a particularly dreadful dimension when it is known that, despite four decades of warnings about the Drins, scientific warnings which commenced as early as the 1950s, Shell only ceased production in the 1980s. Scientific inspection of the evacuated site delivered the most damning of indictments: the Rocky Mountain site was now officially among the most contaminated places on the planet.

Panicked by such a denunciation, Shell made desperate attempts to breathe life back into its barn-owl-loving, countryside-guidebook-sponsoring, kudu-caring, greener-than-thou image by announcing its intention to develop the Rocky Mountain site as a nature reserve. Such an announcement proved to many that the company's high-octane brand of cynicism, allied to its all too apparent belief in the gullibility of the public, had reached an all-time low.

Most certainly, its disastrous record at Rocky Mountain did not prevent Shell from seeking to recruit further profit from the sale of its Drins. While the company had, under official pressure, ceased producing Eldrin in 1982, Dieldrin in 1987 and Aldrin in 1990, it only stopped selling these highly toxic chemicals in 1991, when they were banned from use by the United States government. It was at that juncture that Shell set its sights on countries in the Third World, to where it shipped its remaining Drin stocks.

It can only be a source of amazement that Shell continued to believe for so long that the truth of the matter would not come out and the facts of what was so fast becoming an issue of international concern – a slow but terribly sure impairment of the environment with an awesome potential to visit degradation upon all humanity – would not become a banner to which so many would rally. But rally they did and, indeed, continue to do so.

AN EARLY WARNING

There is, today, very little doubt that the publication of Rachel Carson's *Silent Spring* in 1962 was a principal inspiration and a starting pistol for people right across the face of the earth to begin the race to save the planet from the very worst excesses of a holocaust of the natural order through chemical and industrial pollution. The pioneers of such movements as Friends of the Earth and Greenpeace, united in provocative bravery by the equally brave Rachel Carson, were people who were not prepared to permit their respective governments' dishonest connivance with industrial and commercial practices – which were poisoning the air we breathe and the ground beneath our feet, and laying low the fauna and flora which play such a crucially important role in the delicate, vulnerable balance of nature – to continue without being comprehensively and robustly challenged. In short, they were not prepared to be intimidated by powerful politicians and the cunning of conglomerates, however important both parties considered themselves to be. The clarion call was pure and simple, and had a message all could well understand and relate to: it's our environment as well and we are not prepared to stand by and permit a catalogue of destruction which will bequeath to following generations a shop-soiled, burnt-out, second-hand world.

Such developments put Shell very much on the defensive and the company diverted much of its energy and finances on a counter-claim campaign, the burden of which was, 'Well, this may in certain, isolated instances be happening, but it's nothing to do with us. We are not responsible for this situation.' It was, and, in the opinion of many, remains, a defensive posture that – while displaying a lamentable regard for the truth – the company continues to maintain. Most certainly, the company's history of seeking to suppress the truth of its environmental record in Africa, particularly in Nigeria, as well as America and Europe is, according to its legion of internationally represented critics, as long as it is depressing. Theories abound as to the genesis of such defensive postures as Shell has adopted in recent years, theories which many claim stem from a basic, primitive instinct to protect the company's name and reputation.

On occasions, doubtful presentation of facts and figures by the environmentalist organisations bent on discrediting Shell may well have goaded the company into a style of counter-claims which many regarded, ironically given Shell's own record of such practices, as being little more than smear tactics. That the company had to act against its accusers none would seriously doubt, given that they had, irrespective of whether or not the means which they had applied had been 'fair' or, indeed, 'foul', inflicted serious damage in the public domain on Shell.

In the summer of 1995, environmentalists drew the public's attention to the fact that Shell was attempting to sink one of its redundant oil platforms, the Brent Spar, in the depths of the North Sea. This, said the campaigners, would be a deep-sea time bomb with possibly devastating consequences for marine life and, thus, a direct threat to the fishing communities of the Scandinavian countries and Britain. They said that Shell had no alternative other than to dismantle the platform under environmentally controlled conditions, instead of sending it to the bottom, where it would become a 'ticking time bomb' in the stormy conditions so common in the North Sea.

On this occasion, it did not take long for the captains on the bridge of Shell's ship of state to get the message, and following successive nights of film of the Brent Spar being towed around and through various waters in search of a suitably deep graveyard taking pride of place on television news bulletins, the company capitulated, abandoned its plans to send the redundant platform to the bottom of the sea and, just as the environmental campaigners had asked, made arrangements for the platform to be dismantled under controlled conditions. It was, of course, an exercise considerably more costly than sinking it beneath the waves and in public relations terms the whole event had been a complete disaster, a catastrophe no less.

This was a circumstance Shell would not, could not, let pass without mounting a robust challenge, particularly given the conviction that its critics had been quite unscrupulous in their methods. And so Shell, in effect, took to the barricades, with, it has to be recorded, some considerable degree of success. Indeed, Greenpeace was eventually obliged to issue a public apology which confirmed that it had been quite mistaken all along in the detail it had published which had challenged the technical and scientific data given by Shell in support of its disposal plans.

It was, by any calculation based on fair play, a most considerable triumph for the company but, as with mud, oil sticks and the damage had been well and truly done, with the inevitable result that Shell has continued to find itself on the defensive, comprehensively condemned in the public mind as a master of pollution.

That relatively small and, in a great many instances, poorly funded organisations, which rely principally upon donations from their members and sympathisers among the general populace, could inflict such harm on an international commercial giant such as Shell was in no small measure an indication that the environmentalists had 'come of age'. Most certainly, they had developed an ability to use the media for their own ends and, what is more, were now highly skilled in what many professional journalists saw as manipulative devices.

A particular example of the growing media skill now being widely practised by, for example, Greenpeace, was the use of what is known in the world of television journalism as Video News Releases (VNRs). The organisation made considerable use of these in its highly publicised campaign against the sinking of the Brent Spar. As the saga unfolded, television viewers around the world were treated to nightly scenes of Greenpeace ships and rubber dinghies ploughing bravely through the waves in an attempt to harry the tugs commissioned by Shell to tow the towering piece of equipment to its proposed burial site.

The videos of such heroic and potentially dangerous manoeuvres were supplied to the television stations by Greenpeace, as they had done when the organisation had mounted similar forays against the whaling ships of Norway and Japan. The scenes were designed to pull at the heartstrings and as such recruited an ever-growing army of sympathisers to the environmentalists' crusading banner. Greenpeace was harnessing the power of television and using it to influence public opinion on an international scale.

It was a clever strategy but one with which many practitioners within the journalistic profession were far from happy. The negative aspects of a television newsroom using VNRs supplied by either environmentalist organisations or corporate commercial bodies are self-evident. Such packaged reports are, of course, edited to give the viewer a particular impression that suits the agenda of the organisation that has made them, with no challenging questions being put by a journalist as to the nature and veracity of what is being presented. And even if the news editor decides not to put the VNR to air, its content does strongly suggest the manner in which a 'home-grown' report can be assembled.

It was Greenpeace's extensive use of VNRs depicting its harassment of Shell's Brent Spar operation that led to the practice being called openly into question. Writing in the profession's magazine *Journalist* on the saga, Granville Williams commented on the manipulative nature of VNRs and, quoting a BBC television news editor on the fact that Greenpeace was able to supply 'better, more compelling and more frequent footage than we can ourselves', added, 'Greenpeace exploit our thirst for a good story and for dramatic pictures, and they play on the traditional news values of conflict and confrontation.'

None of which is to suggest, of course, that Shell itself is either an innocent abroad or that it has not pressed into service the dark arts of manipulation and carefully filleted selective detail with which to wage war in the media against its many and varied critics. There is, of course, evidence in abundance of Shell's keen awareness of the essential need to nurse and allay public fears and concerns, and, particularly, to dilute the anxieties of both individuals and groups who campaign on environmental issues. Indeed, the classic route for companies such as Shell in their attempts to convince the public that they care about the environment is to act as benefactor to those agencies which campaign on environmental issues, as its many and much trumpeted donations to organisations ostensibly committed to the mantra of 'green and clean' bear witness. It is reliably believed that the total annual amount given by Shell to such organisations currently stands at about £250,000, a figure of little significance given the gargantuan nature of the company's overall yearly budget but sufficient to illustrate the crucial importance Shell attaches to such an aspect of public relations and its belief that those who condemn its environmental record have to be countered.

As the all too often cash-starved groups have come to learn, such patronage does not necessarily mean that this apparent 'care and concern' will lead to the company making fundamental changes to its business practices which are responsible for the environmental degradation the groups exist to combat. Neither should it ever be assumed that such patronage necessarily means that the company making the donation shares the objectives of the environmental group in question.

It is in the dark arts of attempting to manipulate public opinion that, with the expensive guidance of an assortment of public relations soothsayers, Shell has continued to put its money where its interests are seen to lie. Take, as further example of the 'Now you see me, now you don't' campaigns by Shell designed for it to be seen as both respectable and responsible in the world of environmental protection, the United States-based National Wetlands Coalition. Its logo could not be more eco-friendly – a mallard duck in glorious full flight, soaring above the reedbeds of a pristine wetland. But, upon closer inspection, the coalition is not at all what it strives to appear, its aims a clear contradiction of both its title and its logo.

The Wetlands Coalition was born in the wake of a 1989 statement by President George Bush Snr that it was the policy of his administration to have no net loss of America's wetlands. The prompt establishment of the coalition was the result of a knee-jerk reaction to this presidential declaration of environmental policy, with its financing being met by an alliance of oilmen from Exxon, Mobil and, yes, Shell. Its intention was clear and a direct contradiction of the President's stated aim; in short the National Wetlands Coalition was formed with the covert purpose of maintaining the rights of its members to construct oil installations in the country's wetlands and, indeed, to continue to drill in those very same wetlands without any official impediment whatsoever. The Orwellian instinct and spirit is, indeed, alive and well at Shell! The Wetlands Coalition episode, in telling the tale of the weird and wondrous world of Shell, brings into relief the high-definition contrast between its corporate claims and environmental reality, a constant at the very heart of Shell for many decades.

As many PR professionals regard Shell as having been responsible for some of the most effective campaigns ever mounted, it is of particular astonishment, and to many total bewilderment, that the company has, on so many occasions in recent years, scored so many 'own goals', shot itself so effectively in the foot and, in the case of the reserves crisis of January 2004, shot its shareholders in the back. Indeed, for a company that has long known the true, lasting value of good public relations, allied to its understanding of the long-term damage that can be inflicted by adverse publicity, recent ludicrous failures to take account of such aspects of company governance, from the manner in which it handled the crisis of its own creation in Nigeria to the Brent Spar fiasco and the monumental blunder, the consequence of a connivance that shocked the world, of deliberately overestimating its recoverable oil reserves, such conduct and malpractice virtually defy rational belief or explanation. A senior oil industry insider comments:

> One can only believe that, with the reserves device, it constituted a moment of corporate madness, made possible by the delusion that the company was too big not to get away with it. The whole wretched incident smacks, of course, of an overweening arrogance, an illusory belief in its own propaganda, that it was, simply put, too important an organisation to be publicly called into question.

If this is indeed so, then it illustrates all too clearly that Shell's corporate culture has yet to come to terms with the contemporary world; that what is now known as 'spin' can only take an organisation so far; that there are definite limits on its shelf-life; that in an age of the information highway as epitomised by the Internet, it has become increasingly difficult for an organisation, be it a government or a corporate giant, to prevent damaging leaks, particularly when such detail lays bare information on company activities it has sought to conceal from public gaze. And if there is a moral here, a lesson to be learnt, it is surely this: that after years of spin, smokescreens and mirrors, posturing and a lamentable catalogue of a suppression of facts – stubborn things, facts – underwritten by unrelenting PR campaigns, the cold stark truth for Shell is that it is its corporate culture, a culture which made all of this possible, that, far from being the solution to its ills is, in actual fact, the problem.

CHAPTER 9
DIRTY WORK

By the mid-1980s, alarm over Shell's international operations was spreading with the speed of a bushfire and particularly among the world's few remaining aboriginal communities, which have been especially vulnerable to the company's freebooting method of conducting the business of its commercial empire. From the frozen wastes of Alaska and Canada's vast tundra, the voices of the Inuit people were raised in protest at the designs Shell had for their native lands. They were far from being alone. In Australia, the long-suffering Aborigines protested at Shell's operations in their sacred ancestral lands, as did the indigenous people of Brazil, the Indians of the mighty Amazon Basin, upon whom the twentieth century wrought such death and destruction.

The times they were, indeed, a-changing and with an abruptness that is known to have shocked many in Shell: the company was placed 'on charge' as never before. Throughout the 1980s, demands, on an ever-increasing international scale, rained down on Shell: demands that the company not only cease its operations in some of the world's most vulnerable lands but also that it make recompense in full measure for the damage and destruction it had caused both to lives and to the environment. One such demand came from the fastness of the Peruvian rainforest, in which Shell had carried out exploration activities. As a direct result of the presence of Shell personnel in Peru's rainforests, some 100 Nahua Indians perished after contracting diseases against which they had no natural immunity. On this occasion, Shell took evasive action, protesting the company's innocence and pointing its corporate finger of accusation at commercial loggers, who, the company claimed, had operated in the Nahua's tribal lands before they arrived on the scene.

But the rising tide of anger now closing over Shell's head was not confined to the protests and demands for restitution being made by the indigenous peoples in the frozen wastes of the world's far northern lands, the Aborigines of Australia's baking deserts and the last of those to be felled by the fatal impact of Western man and his pursuit of profit at all costs, the Indians of the South American rainforests. By the close of the 1980s, the company's carefully crafted image of being a socially and environmentally 'good neighbour' to the world and his wife, and the impression that its blue-chip reliability could, well, be relied upon, was also coming under public assault in both the United States and the United Kingdom.

In April of 1988, 440,000 gallons of oil was discharged into San Francisco Bay from Shell's Martinez refinery, as a result of which thousands of birds perished in the now heavily polluted, oil-laced waters of the Pacific. In 1989, the company was responsible for an oil spill into the River Mersey, some 150 tons of thick crude. Shell was fined £1 million, a record amount to be imposed by a British court of law for environmental damage.

Come 1990, the board of Shell at last realised that it was now obliged to make a response, beyond its customary public relations-inspired strategies, to the rapidly growing global climate of environmental awareness. Speaking in 1990, Sir John Collins, then Shell's chairman, said, 'The biggest challenge facing the energy industry is the global environment and global warming. The possible consequences of manmade global warming are so worrying that concerted international action is clearly called for.'

For a company which had, hitherto, steadfastly refused to own up to any environmental transgressions and had denied outright that its operations were responsible for a catalogue of calamities which, to others, were so clearly born of Shell's activities, it was a statement of future responsible intent that was little short of revolutionary and, indeed, there were many who, at the time, saw Sir John's statement as bordering on a Damascene conversion by Shell.

Hope does, indeed, spring eternal and Shell watchers waited for the company to take initiatives that would support public anticipation that, at last, a whole new approach to the manner in which it conducted its business, from exploration in virgin lands with finely balanced environments to the extraction of oil from the bowels of the earth, its refining and, ultimately, its carriage across the world's seas and oceans, was in the offing. For, just as the company chairman had said in his groundbreaking speech, was not the problem of man-induced global warming one that called for concerted international action?

It did not take long for such high hopes to be cruelly dashed on the jagged rocks of Shell's age-old practice of the creed of enlightened self-interest. For despite Sir John Collins' speech, the company joined the Global Climate Coalition, which had spent millions of dollars in an attempt to influence the United Nations' climate negotiations.

The Kyoto conference on global warming, which was attended by countries both great and small, and which concluded in December 1997, was, throughout its deliberations, subjected to the constant lobbying of the coalition's public relations gurus, one of whom, in a clear contradiction of the spirit of the Shell chairman's remarks on global warming, declared at Kyoto, 'There is no clear scientific consensus that man-induced climate change is happening now', a statement which came two years after the world's leading scientists had agreed that there was. Such a pronouncement assumes an even more dubious dimension when it becomes known that even given its financial support of an organisation established to dismiss scientific warnings on global climate change and with it a rise in the levels of the world's seas and oceans, Shell ordered that the height of its Troll platform in the North Sea be increased by one metre.

DODGING THE DOWNSIDES

Shell's practice of saying one thing whilst quietly doing another is not confined to one issue, country or, indeed, to one continent. It is a comprehensive practice of which the company's conduct in the oil-producing state of Texas serves as just one example.

The Motiva Refinery at Port Arthur in Texas, a joint venture in which Shell is the majority shareholder, has long been regarded by local people as the biggest single serial polluter in the country and the single biggest environmental offender, so much so that local residents have long referred to it as 'the neighbour from Hell', with just and painful cause. Motiva is a vast refineries complex, covering some 3,500 acres and, as the installations have continued to expand, some are now situated in close proximity to people's homes with, in many cases, only a chain-link fence separating them (the area has long been known as 'Gasoline Alley') from the backyards of many sick, angry and deeply unhappy people.

Hilton Kelley is a Port Arthur man who moved to Hollywood to pursue a successful career as a stuntman, but so angry and, indeed, alarmed did he become at the continuing levels of pollution spawned from the refineries that he abandoned his place on the silver screen to return home and take up environmental cudgels on his community's behalf.

Speaking on BBC Radio's *File on 4* programme on 23 March 2004, Kelley said:

> The refineries are now in people's backyards. Often the air here smells of rotten eggs or even more strongly of ammonia, or there's a kind of sweet odour from gases being released into the atmosphere, which, when mixed, become something else.

Indeed, so alarmed did he and his fellow residents become about the quality of air they are obliged to breathe that the US Environmental Protection Agency (EPA), in response to the pleas of the people of Port Arthur, led by Hilton Kelley, established a mobile laboratory in the town, the principal feature of which is a highly sophisticated Trace Atmosphere Gas Analyser (TAGA). As the laboratory was driven around Port Arthur, the TAGA system was used to identify some 13 known hazardous chemicals and pollutants.

Wilma Subra, an award-winning analytical chemist of international stature, conducted a detailed analysis of the tests made by the TAGA and was deeply disturbed at what she found. Also speaking on *File on 4*, she said:

> The highest concentration was of benzine, a known cancer-causing agent. This is the chemical most frequently released by refineries such as Motiva. Benzine was present in concentrations of up to 175 parts per billion, which far exceeds the accepted level set by the State of Texas. On an annual basis, the State

> permits 1 part per billion. So these hits, recorded as the TAGA truck went up and down the roads and streets of Port Arthur, were up to 175 times the accepted limits.

The distinguished analytical chemist continued:

> Motiva releases 2,200 pounds of benzine into the air over Port Arthur in a year. This is a cause for particular concern because chemicals released by Motiva exceed the officially permitted levels of pollution and constantly expose residents to substances such as benzine.

But it's not just the high levels of benzine in the air over Port Arthur that local people have to worry about. In Gasoline Alley, there are other chemicals in the air which when combined constitute what Wilma Subra described as 'a toxic soup'.

Shell's head of sustainable development at Motiva – the company affects not to see the irony of such a designation – Bill Wimberley is keen to point out to visiting investigative journalists two recently installed and sophisticated pieces of anti-pollution equipment, a wet gas scrubber and a flare gas recovery system, installations which, he claims on Shell's behalf, not only make the Port Arthur air safe to breathe but are actually improving its quality. When put under pressure to substantiate such claims, Wimberley is on record as commenting, 'I do not dispute the TAGA findings but I might disagree with the conclusions.'

Local people, however, exercise no such ambiguities. Indeed, they complain with more than a trace of bitterness that pollution levels are so high because in addition to the daily, routine emissions from the refineries, there is a high incidence of accidents or, as Shell prefers to style them, 'upsets'. Similarly, the company never uses the term 'dangerous', resorting to the less direct 'hazardous'. However, on the essential burden of the complaint, there is no legitimate room for either ambiguity or evasion behind the use of differing terms. When any part of the plant has to be shut down during 'upsets', chemicals are vented to chimneys or burnt off in flares, meaning, of course, that the chemicals are released into the atmosphere over Port Arthur. Indeed, often on a daily basis, the clouds of toxic emissions are clearly visible over the refineries, with the accompanying noxious smell leaving no doubt as to their harmful effects. Wilma Subra has carefully analysed the information Motiva is obliged to file with the state of Texas regulator when such 'upsets' occur. Commented this distinguished professional:

> Motiva is reporting five, ten and, on occasions, twenty 'upsets' each day. During the first two months of 2000, over a two- to three-day period, there were excess emissions in the west of the Port Arthur area due to accidental releases and 'upsets'. Sometimes there are 30, even 40, 'upsets' in a single day, resulting in chemicals being released into the town's atmosphere.

In more recent years, Motiva claims that it has significantly reduced the number of 'upsets' but in Texas, where oil continues to be king, companies are responsible for self-reporting to the regulator: in short, they are allowed to police their own activities. As a result, the company has been fined for failing to report all of its excess emissions at Port Arthur and in 2002, fines were imposed in respect of 37 incidents related to excessive emissions and/or failing to report to the regulator 'in a proper manner'.

Bill Wimberley did make attempts, yet again, to contest the fine detail of his employer's record in the matter of 'upsets' and failure to report emissions of noxious chemicals into the Port Arthur atmosphere that were in excess of levels set by the state of Texas regulator, but, in the event, met with little success. Indeed, he was eventually obliged by the very stubborn nature of the facts of the matter to confirm that between June of 2002 and July of 2003 there had been some 42 notifiable breaches of state regulations by Motiva. These were breaches for which local people were paying an increasingly heavy price, with many of their children, 13 in the western area of Port Arthur alone, having been diagnosed as asthmatic, their condition being so serious that it could only be temporarily alleviated by their use of nebulisers.

Oil might, indeed, continue to be king in the state, but with public health in Port Arthur so obviously being grievously affected by the Motiva refineries something clearly had to be done. The first troops into the battle lines being drawn between the Port Arthur community and the company were a team of toxicologists from the University of Texas's medical faculty. They were armed with a symptom survey in which local people were asked to give details of the condition of their health. Such details were then compared to results obtained from an identical survey with the same demographics located 75 miles distant, at Galveston, where the refining of crude oil is conducted well away from residential areas. Dr Deborah Morris, a genetic toxicologist, reported that the most disturbing aspect established by the Port Arthur survey was that it showed a very high incidence of respiratory problems, such as asthma in children and a range of other respiratory ailments, from persistent coughing to lung disease, adding:

> For the survey, we only considered non-smokers. Overall, the survey revealed that 9 per cent of the people in the Galveston community complained of health problems, whereas in Port Arthur the figure was 70 per cent, a substantial difference and a very high number indeed.

Dr Morris concluded, 'I have absolutely no doubt whatsoever that air quality was the major factor in the results of the survey.'

Yet again, Shell's faithful Mr Wimberley was at hand to offer alternative and, on this occasion, incredible reasons for the results of the survey. The real culprit for the respiratory problems among the people of Port Arthur was, claimed Wimberley, pollen. And if not pollen, well, dust. And maybe it was both pollen and dust. Motiva meanwhile continues to formally deny that its refineries are causing health problems through toxic emissions while, simultaneously, being obliged to confirm that, together with five others in the complex, it is facing lawsuits from, quite literally, hundreds of people in the Port Arthur community.

Such a failure to accept accountability at the highest of levels within Shell's management should really come as no surprise according to the senior oil analyst Greg Muttit. 'When the company announced its extremely clever "Profits and Principles" blueprint in 1998, it also put in place, with no public fanfare at all, a programme of business imperatives which are wholly and totally at odds with the ballyhoo strategy of "greenness" and "environmental concern".'

Muttit identifies the principal business imperatives, of which there are three. 'Personal accountability', which, contrary to the impression the term is undoubtedly designed to convey, has actually meant a much greater degree of centralised control. 'Cost control', the burden of which was, and indeed remains, plainly obvious and, third, the ominous-sounding 'Capital discipline'. This last imperative translates into Shell being unwilling at best and resolutely opposed almost all of the time to any nature of investment which might, even temporarily, diminish the ultimate bottom-line benefit. All three imperatives are clearly in conflict with the crafted public image of a listening, caring corporation and they are, with equal clarity, conflicts which Shell has failed to reconcile or, indeed, make any serious attempt to do so. And yet the litany continues, with the company telling the world that it really has cleaned up its act, and losing no opportunity to trumpet its social and environmental good-neighbour credentials in tandem with yet further assurances of its continuing blue-chip reliability, all of which, in the face of proven, stubborn facts, can but raise the question in the public mind: 'Can we really be sure of Shell?'

As the legal fight in Texas continues, one of the most disturbing aspects of Shell's record in the state is this: if the company is prepared to behave in such a socially irresponsible manner in the Land of the Free, deep in Crockett country and on the home turf of litigation-crazed, predatory lawyers, just what do they consider themselves free to do in states of the Third World, where lax, poorly applied laws are all too much in evidence and where corruption among officials is exercised on a truly horrific scale? What, for example, has Shell got up to in Africa's most populous nation, that giant of a land on the continent's western shore, the republic of Nigeria?

CHAPTER 10

DEATH ON THE DELTA

On 20 July 2004, Shell made an announcement which delighted many and was the cause of considerable surprise to others, particularly so given that it appeared to scotch previously held, well-founded rumours that it was on the verge of quitting Nigeria, where it has long been the country's biggest oil producer. The announcement confirmed the appointment, for the very first time, of a Nigerian to be head of its operations in the country. Basil Omiyi, 58, joined Shell in 1970 and his appointment as Shell's managing director in Nigeria met with a well-justified chorus of approval, particularly from the country's oil-workers' unions, which have long been applying pressure on Shell to promote more local people to the top jobs in the company. 'I look forward,' said Omiyi, 'to the opportunities and challenges which lie ahead.' To which he could well have added, and there have been more than a few in recent years.

Shell's record in this vast, sprawling African republic, home to numerous bloody coups and counter-coups since it achieved independence from Britain over 40 years ago and an international byword for staggering state incompetence and wholesale corruption, is a far from happy one. The country is a member of the Organisation of Petroleum Exporting Countries (OPEC) and is the world's seventh largest exporter of crude oil. Yet, as recent events have shown, its exports are not always of the traditional kind.

On 24 October 2003, the London-based *Guardian* newspaper ran a story from its Africa correspondent, Rory Carroll, which told a tale of twenty-first-century high-seas smuggling, the latest revelation in a long, bloody catalogue of crimes spawned by Nigeria's oil industry. Carroll's report confirmed what many in Nigeria's oilfields had known for a considerable number of years: that criminal gangs were siphoning off huge amounts of crude from the oil pipelines in the Niger Delta. But the *Guardian* story brought a new dimension to this particular tale of wholesale theft and the method of disposal of the black booty. The report continued:

> A Russian-registered tanker laden with 11,300 tons of allegedly stolen crude has become the latest vessel to be intercepted by the Nigerian Navy in the Gulf of Guinea. The vessel, the *African Pride*, is believed to be part of a fleet which aids the theft of an estimated 200,000 barrels a day from the Delta's swamps. The tanker had the biggest consignment of all the 15 vessels seized since January, said Antonio Ibinabo Bob-Manuel, a Nigerian Rear-Admiral. Its crew of 18 Russians, two Romanians and two Georgians are in jail awaiting a court hearing.

Speaking at a press conference, Rear-Admiral Bob-Manuel said the thieves were scaling up from barges to tankers, with each cargo of stolen oil worth at least $10 million. The ships were said to have been intercepted in areas around Nigeria.

What started off with a few amateurs wrenching open a pipeline to extract oil to sell on the local market has in recent years grown into a vast criminal enterprise that swallows 10–15 per cent of Nigeria's daily output of 2.2 million barrels, according to officials. Siphoning off such quantities in a landscape of swamp and jungle with thousands of creeks requires the most sophisticated equipment and organisation, and to the dismay of the Nigerian government and, indeed, the oil companies, the thieves have proved that they have both in large measure. But when the thieves' plans go awry, they can prove fatal.

The Reuters news service reported during the very week of the *Guardian* article that six people had been burnt to death at Port Harcourt, in the Delta, when thieves ignited a fire while scooping up fuel from boats. This is not an isolated case, as it has been estimated that during the past five years hundreds of people have died in similar circumstances. The government of Nigeria has accused rival and ethnically constituted armed groups, especially young men from the Ijaw tribe, of being responsible for the widespread theft and consequent death and destruction.

But the London-based president of the Ijaw Association claims that his people are not the culprits and that senior political and military figures in the Nigerian establishment are to blame. Comments Rowland Ekperi, 'They are trying to blame the local people but they don't have the heavy-duty trucks and ships for this sort of thing. It's members of Nigeria's elite who are no longer in power who are responsible. They use money raised from the sale of the crude oil to finance their political activities', adding that the crude was rumoured to end up in nearby states such as the Ivory Coast, Benin and Gabon, as well as Germany.

What is not in dispute is that, irrespective of the ultimate destination of the stolen crude, attempts to crack down on the smugglers have met with little success, further evidence, say many senior Nigerian officials, from both within the country and beyond, that the federal government and the military establishment are, indeed, involved in the racket and that cash raised from the sale of the smuggled oil finds its way into their pockets. The same former senior officials also say that what is making such a sorry situation even worse is that the oil companies continue to plunder the country's resources, while abandoning the people of the Delta to poverty, unemployment, pollution and subsequent disease.

It was not, however, the increasingly sorry condition that so many in the oil-producing areas of the Delta found themselves in as a direct result of having, quite literally, the oil industry operating in their very own backyards that attracted the concern of the countries of Western Europe, the United States or the international community at large. Indeed, little action was taken on the smugglers' activities until the oil companies themselves began to apply pressure on Washington to do something about a situation that the Nigerian government, even given the very best efforts of its small navy, seemed largely unable to control.

In the event, America did take measures to help the world's seventh largest exporter of crude oil by donating three fifty-six-metre, refitted, Second World War patrol boats for anti-smuggling use by the Nigerian Navy, with a promise of four more at a later date. The Pentagon financed each vessel's rehabilitation to the sum of $3.5 million.

What success the navy will now have against the smugglers remains to be seen, but prior to 2004, it was more commonly the involvement of Nigeria's army in the Delta, particularly its use of brutal tactics against its own people and the role of senior military figures in corrupt practices leading to their own personal enrichment in the smuggling operation, which commanded world headlines.

THE HAND OF HISTORY

In late March of 2003, within days of the commencement of the second Gulf War, the Anglo-American attack and invasion of Iraq (which caused international concern about future supplies of oil and had sent the price soaring back to the 'oil shock' years of the 1970s and 1980s – $40 a barrel), the Nigerian Army claimed that its troops had entered the Delta region in order to bring it back under government control after more than a week of sporadic violence. Speaking on 24 March 2003, the army's spokesman, Colonel Onwuamaegbu Chukwuemeka, told members of the press that a military task force had acted to prevent Ijaw militants mounting further attacks on oil installations and to stop fighting between local rival communities.

That such attacks were taking place is in no doubt. Indeed, three of the principal oil companies working in the Delta had been obliged, because of the state of unrest, to cut back on their activities, with the combined loss of production representing one-third of the country's usual output. ChevronTexaco shut down its main export terminal, Total Fina Elf had pulled out of an oil-storage depot and Shell had withdrawn from four of its facilities.

The Nigerian Army has never enjoyed a reputation for 'softly-softly' tactics and within days it confirmed that 13 people had been shot dead in the violence, a number which was immediately disputed, with a local human-rights activist, Danka Pueba, telling the French News Agency AFP that the death toll among the local population as a result of military activity was far higher than the number given by the army. Most certainly, events on the ground support such an assertion, with thousands of Ijaw refugees fleeing in fear of the soldiers, claiming that their community was in a virtual state of siege from both land and sea, with the army blockading their villages while they were being shelled from navy gunboats out at sea. The army, meanwhile, continued its claims that it was not attacking peaceful civilians and stressed that it was using minimum force.

The Ijaw, however, remained defiant, with the more militant leaders threatening to detonate oil installations and make the area ungovernable if the army did not end its armed incursions into their villages. They also demanded a greater share of the oil wealth being generated from their traditional lands.

Oil was the genesis of such civil strife and in the dark, dank swamps, creeks and tangled undergrowth of Nigeria's multi-tribal Delta, it has often been the spark which has ignited the flames of dissent, despoliation and social and political anarchy, a situation to which, then as now, thrusting foreign interests have made a particularly combustible contribution.

In the early 1890s, the British and the Portuguese, both of whom had long traded with the tribes of the Delta, initially in slaves for their respective colonies and for the plantations of America, turned their attention to raw materials, primary products so sorely needed at home, where they became essential ingredients in the expansion of the industrial revolution and in the development of a broad range of manufactures.

Principal in such a range of native produce was not today's crude oil, sucked from the bowels of the Delta by means of advanced technology, but palm oil, produced from the nuts of the palms which all too often rise out of pools of stagnant crude oil and which are such a feature of the region's landscape. When boiled, the nuts surrendered a thick, rich oil ideal for the lubrication of machinery and, when refined, for the production of a wide range of consumer goods, such as toilet soap. It was, for both the traders from an alien world and the native people of the Delta, a mutually beneficial trading arrangement, and within a short timespan, the region became the single biggest source of the commodity on the African continent, with the trade being virtually controlled by traders from the port of Liverpool and the Delta chiefs, those astute, hard men who had waxed rich on the supply of their own people to the slavers on board their ships at anchor in the bay. They also displayed a sharp business acumen in the new trade of palm oil, preventing personal contact between the Europeans and those who actually produced the oil.

When the pioneering European traders first arrived in the Delta, the dominant tribe in the business of selling both slaves and palm oil was the Ijaw, whose traditional occupation centred on small-scale subsistence cultivation and fishing. While, for wholly understandable reasons, as we shall see, the world beyond Nigeria's shores associates another Delta tribe, the Ogoni, with resistance to the oil companies in the marshalling of public dissent against the activities of the Western oil companies, by the time Shell confirmed its discovery of oil in the late 1950s in their tribal territory, the Ijaw, together with other minority tribes in the Delta, had long smarted under what was perceived to be a central government in Lagos which not only exercised too much power in the region but also treated local tribes as second-class citizens. Now, of course, with the discovery of oil, a resource upon which the economy of the country would in future turn, the government was going to exert an even tighter hold over the region and dismiss, with all the force at its command, any attempts by the Ijaw, or, indeed, any other tribal group in the Delta, to secure for themselves any degree of political autonomy.

Thus did it fall to an Ijaw radical, Isaac Boro, with endorsement by fellow dissidents, to issue a challenge not just to Lagos but also to the recently arrived oil companies, doing so by establishing a community-based organisation, Integral WXYZ. One of the organisation's premier initiatives was the formation, as an extension of Integral WXYZ, of what was termed the Niger Delta Council, which threw down the gauntlet to the oil companies, and to Shell in particular. One of the council's first published demands, in the early 1960s, was a direct and explicit challenge to Shell.

Writing in his *The Twelve Day Revolution*, published in 1982 after his death, Boro recalled that the council took issue with the oil giants over 'their continued atrocities to our people and their wicked reluctance to improve the lot of the people they were to be associated with for long'. The council also made out estimated bills for payment by the oil companies with respect to inadequate damages paid to natives for cash crops and economic trees destroyed during their operations.

It was a demand to which Shell made neither reply nor response, and to the anger of the Ijaw and of other tribal communities in the Delta the state of affairs continued, particularly the widespread pollution by crude oil of both the fertile soil and the waters of the region, upon which local people had for so long depended.

But it was armed insurrection on 15 January 1966, against the government of Sir Abubakar Tafawa Balewa, which swept aside the peaceful existence of Integral WXYZ and its Niger Delta Council. Sir Abubakar, who had led Nigeria to independence from Britain in 1960, and who was held in particularly high regard by the Ijaw because his administration had decreed that the Delta was to be afforded special status within the country's federal structure, was brutally killed by army officers, ironically most of whom were Ibo from the Delta, with his body later being found in a ditch.

Oil, or the corrosive effect of the easy money it generated, had struck again, as one of the principal reasons for the armed coup by the young Ibo military officers was an articulation of the widespread public anger with those in the government termed 'the ten-per-centers', ministers and senior state officials who were creaming off such amounts from contracts and the hitherto unparalleled revenues arising from the gushing crude.

The coup electrified Boro and his followers. As he recorded in his book, 'The only protector of the Ijaw, Sir Abubakar, had gone. If we did not move now, we would throw ourselves into perpetual slavery.'

Boro was a man of undoubted resolve and within days of the coup, he and his Ijaw allies had raised, with the minimum of finance – about £150 – the banner of armed resistance, a red flag with a crocodile at its centre. Named the Niger Delta Force, initially just 160 in strength, the group was not long in taking political action and in February declared the Delta region an independent republic and announced a state of emergency to govern it during its infancy.

It was a development that the new military government in Lagos met with prompt force, principally to deter other tribal groups from emulating their Delta cousins but also, of pivotal importance, to shore up Shell, which was in a state of high panic and alarm at such a disruptive turn of events. Indeed, so alarmed was the company that it quickly resorted to a practice both tried and tested: it associated itself with the military action, as it had done elsewhere when its interests were threatened. The company supplied the Nigerian army with a flotilla of pontoon boats with which to attack Ijaw positions. In short, Shell, true to its tradition of taking a discreet position on the path of political power in countries in which it operates or where it seeks to be in a position to exert a behind-the-scenes influence at the executive level, threw its not inconsiderable weight behind a military establishment which had seized the commanding heights of government via the unconstitutional expedient of high treason, murder and a usurping of an administration which, for all its manifold faults, had achieved its rule through the ballot box.

Isaac Boro's movement, for all its bravado, was, of course, no match against such force, an army of ruthless intent backed by one of the world's most powerful companies. The end could not be long in coming for the Niger Delta Force and it arrived soon enough. At the end of the first week of March, Boro surrendered to officers of the Nigerian Army who placed him and his followers on treason charges. After a perfunctory trial, they were sentenced to death, sentences which were later commuted.

Ironically, Isaac Boro went on to fight on the federal side in the 1967–70 Biafran War, a conflict which was itself ignited by oil when the Ibos of the Delta sought to secede from the Nigerian Federation, confident that the riches from oil were rightfully theirs and that the revenues would sustain their breakaway state of Biafra. The government in Lagos, of course, took a contrary view, as indeed did Shell, with the company's support for the military government in the four-year-long bloody civil war remaining, today, a matter of very considerable bitterness to many in the Ibo establishment. While the death toll has never been accurately established, many thousands died from starvation towards the end of the conflict due to a military blockade of Ibo territory. It is reliably believed to have been one of Africa's most horrendous civil wars, which, given the continent's abysmal record of its governments visiting wholesale death and destruction upon their own people, is saying really rather a lot.

In any event, it was a war in which Isaac Boro lost his life and, for a while at least, the fires of dissent caused by the havoc being wrought by the activities of the oil companies in the Delta were doused, although the embers of resentment have continued to glow and, on occasion, flare into public protest. Indeed, in the 30 years and more which have passed since the end of the Biafran War, the tribal communities of the Delta have continued to regard themselves as a people dispossessed of the revenue being derived from their ancestral lands and have watched with ever increasing anger the continuing pollution and consequent destruction of their native landscape.

A LEADER OF MEN

The leader of the Ijaw may well have perished on the field of battle and his followers been temporarily disheartened, but in 1970, the year which saw the end of the Biafran War, the torch of protest at the conduct of the oil companies in the Delta passed to leaders in the Ogoni tribe, an intervention about which, in the tragic and melancholy event, the world was, in time, to hear more, much, much more.

The tribal area of the Ogoni is, by comparison with the vast geographical area of many other communities in Nigeria, small, very small in fact, comprising just over 400 square miles, with a population of approximately 500,000. Ogoniland is situated in the south-east of the Delta, just east of Port Harcourt, capital of the Delta's Rivers State and the most significant urban area in the whole of the region. Since Shell began operations in Ogoniland, it has extracted a reliably estimated 634 million barrels of oil, with a value of some $30 billion. This is a vast sum by any yardstick but the long-suffering Ogoni people have received little by way of material return. On the contrary, much of their hitherto productive land has suffered quite the most devastating despoliation as a result of crude oil pollution while wholesale poverty has continued unabated. The provision of health and education services remains minimal, with only the fortunate few receiving such basic utility supplies as water and electricity.

While, under a succession of both civil and military governments (during the first 40 years of independence from Britain, the army ruled for 30 of them), a revenue distribution system has been in place, ranging from an initial 1.5 per cent to 13 per cent of income derived from oil-producing areas, the Ogoni have been starved of financial resources with which to establish the supply of public services or to conduct any development projects. In short, successive administrations have failed totally to honour such a revenue-sharing arrangement, a fact which, to be scrupulously fair to Shell, it has, itself, acknowledged but, alas, done little or nothing to put to rights. Indeed, and as the record clearly establishes, it only treats with the government of the day when it is threatened by civil unrest.

As early as 1970, however, senior members of the Ogoni tribal establishment had become so alarmed at the rapidly deteriorating condition of their lands due to oil pollution and at the continuing poverty of their people at a time when their native area was producing so much wealth for others that they wrote a joint letter to the military governor of Rivers State, which read, in part:

> May it please Your Excellency to give your fatherly attention and sympathetic consideration to the complaints of your people of

> Ogoni Division, who have suffered in silence as a direct result of the discovery and exploitation of mineral oil and gas in this Division over the past decades.

The letter continued by requesting a more equitable share of the revenue from the crude oil pumped from their tribal lands and that Shell be obliged to take immediate, remedial action to repair the pollution its activities had caused and pay greater attention in future to the region's environment. It was a just and plaintive cry but one to which the military governor of Rivers State made no reply. What, however, the tribal elders could not have known at the time was the effect their letter was to have on a young man then serving as the Rivers State Commissioner for Education. His name was Ken Saro-Wiwa.

The Ogoni may well be a minority tribe, a small community in a vast land, one inhabited by tribal giants, but what they may lack in numbers they more than adequately make up for in raw courage and audacity. It was an aspect of their nature which gave the British particular problems when, in 1914, they were obliged to dispatch a military mission in order to subdue them. The year 1914 was significant in that it was the year in which Nigeria came into existence as an embryonic nation state, albeit from a colonial womb.

Ken Saro-Wiwa, in keeping with the realities of his own Ogoni people, was a small, slight figure but capable of roaring like a fearless lion, a well-orchestrated sound in tune with a formidable intellect. As such, he inspired both love and loathing, the former from his increasingly emboldened followers, who saw him as their champion against the forces of a corrupt and incapable government and a leviathan in the all-powerful world of business which was raping a high-revenue resource from their ancestral lands while polluting them beyond redemption in the sad and increasingly wretched process.

To the latter, Shell, Saro-Wiwa was an arch troublemaker, a man hungry for political power on his own terms, one who did not hesitate to exploit Ogoni problems for his very own personal ends. A servant of the state he may well have been, a Commissioner for Education no less, but it was as an author, a journalist and an Ogoni political activist that he came to the attention of the world beyond the shores of his own country and eventually to make his ultimately tragic way into the world's headlines.

Saro-Wiwa never deviated from his opinion that the problems of his own people had commenced in 1914 when they were taken into Britannia's all-encompassing embrace. Writing in his political memoir, *A Month and a Day: A Detention Diary*, he delivered a particularly damning indictment of Britain's administrative arrangements for the Ogoni people. 'We were forced into alien administrative structures and herded into the domestic colonialism of Nigeria.'

While such comment is a run-of-the-mill statement from an African nationalist, it does, nonetheless, have a considerable degree of merit, principally a just condemnation of the practice by Europe's colonial powers of both dividing tribes and amalgamating them to suit administrative purposes. However, it does, somewhat disingenuously, choose to ignore the fact that the Ogoni people's troubles had commenced well before the arrival of the British on the scene. Long before the unstoppable forces of imperialism had breached West Africa's shores, Britain's outlawing of slavery had led to an explosion in the Ogoni population, with the inevitable pressure on available land suitable for cultivation. As a direct consequence of such a land hunger, the tribe was obliged to venture ever deeper into the forests of the Delta in search of land to cultivate. The principal method used to clear the forest was the age-old African practice of 'slash and burn', with the result that the thin layer of earth upon which cultivation was possible was rendered unstable and during the wet seasons it was often washed clean away, revealing a thin, sandy soil quite ill suited to growing crops.

It was a process that continued well into the twentieth century, indeed right up to the time of the arrival, in the early 1950s, of the men from Shell. The final act of degradation of so much of their land was at hand, as the oilmen constructed terminals, flow stations, gas-flaring plants and pipelines, which all too often leaked the poisonous crude into their ancestral lands and fouled their creeks and tributaries. And to add high insult to a most grievous injury, the wealth being extracted from their land was denied to them.

The Ogoni, knowing full well that they faced an unholy alliance between a corrupt government in far-off Lagos and a multinational company which gave a very good impression of being quite impervious to the despoliation they were inflicting, looked for a leader who would make their voices heard. In their very own Ken Saro-Wiwa they found a giant among men who was ready for the fight.

There is much of the 'David and Goliath' character about the campaign launched in 1990 by Ken Saro-Wiwa, a campaign that was articulated and taken into the international domain by a small, frail-looking man (he stood just 5 ft 2 in. in his stockinged feet) on behalf of a minority, hitherto little-known, people. Indeed, it says much of his incisive skill at recruiting the attention of an international public that within weeks of the launch of his Movement for the Survival of the Ogoni People (MOSOP), both he, personally, and his cause were well known to presidents, prime ministers, crowned heads of Europe and the myriad of organisations dedicated to promoting the protection of the global environment.

Yet while his campaign generated both heat and light, back home in Nigeria the crisis continued unabated, with the Ogoni people continuing to be denied revenue from the pumping of crude oil from their increasingly polluted lands. But Saro-Wiwa persevered undaunted and in 1990, the same year in which he established MOSOP, his book *On a Darkling Plain* was given a public launch in the Nigerian capital, Lagos.

The launch marked a most significant extension of his fight, as an Ogoni activist, with the government over its manifest failure to return at least some of the revenues from the export of oil to the tribe and with Shell over the degradation of his people's land that its activities had caused, and were continuing to cause, on a quite horrific scale. In short, he took the fight to the very heart of the Nigerian establishment by questioning the nature of his country's creation, its governance and the connivance by national leaders in the environmental damage being inflicted upon the land by multinational companies.

It was, in very real effect, a pioneering call against the very nature of globalisation and as such it echoed around the world. With regard to the inherent complexities attending the country's creation, which, he had long declared, were fundamentally responsible for the gross mismanagement of its national affairs and the comprehensively practised corruption which was its handmaiden ('Corruption in Nigeria is so easy because it is made so easy to practise'), Saro-Wiwa told his audience, 'The present division of the country into a federation in which some ethnic groups are split into several states, whereas other ethnic groups are forced to remain together in a difficult unitary system inimical to the federal culture of the country, is a recipe for dissension and future wars.'

But the cause he had made his own, the dispossession of his Ogoni people of what was rightfully theirs and the destruction of their land by the activities of the oil companies, and by Shell in particular, quickly followed: 'The system of revenue allocation, the development policies of successive federal administrations and the insensitivity of the Nigerian elite have turned the Delta and its environs into an ecological disaster and dehumanised its inhabitants.'

While his speech at the book launch was received with fury by the country's military rulers and with consummate dismay and mounting anger among the men at Shell, it was his writing as a journalist that was principally responsible for Ken Saro-Wiwa making the most serious enemies at very senior levels indeed. An article he had written for Nigeria's *Sunday Times*, in June of 1990, in which he attacked the military governors of Rivers and Cross River states for acts of gross corruption, was, in effect, banned from publication in the wake of pressure on the owners of the newspaper. It was a decision which both dismayed and angered Saro-Wiwa but he was far from done. In what proved to be his final column in the *Sunday Times*, which he titled 'The Coming War in the Delta', he wrote in biting terms of the oil companies' polluting record, while reserving his most potent bile for his old enemy Shell. The article appeared in the paper's first edition but was pulled from the second, on official demand.

The Ogoni people, he wrote, 'are faced by a company, Shell, whose management policies are racist and cruelly stupid, and which is out to exploit and encourage Nigerian ethnocentrism'. He continued by calling upon the country's military rulers to:

> pay royalty to the landlords for oil mined from their land and the revenue allocation formula must be reviewed to emphasise derivation. Citizens from the oil-bearing areas must be represented on the boards of directors of oil companies prospecting for oil in particular areas and communities in the oil-bearing areas should have equity participation in the oil companies operating therein. Finally, the Delta people must be allowed to join in the lucrative sale of crude oil. Only in this way can the cataclysm that is building up in the Delta be avoided.

Such a clarion call ended with the plaintive cry, 'Is anyone listening?'

It may well have been the final foray into journalism permitted him by a military and company establishment now so visibly angered and shaken, but, as events were so quickly to prove, people were, indeed, listening, not least the traditional leaders, the chiefs of the Ogoni, a body of men who did not always see eye to eye with intellectuals within their community. That such a conservative element was prepared to take some sort of action and in doing so acknowledge publicly that something had to be done in order to arrest a rapidly increasing climate of public unrest was, in itself, a not inconsiderable advance for the movement to which Saro-Wiwa was now totally committed.

The chiefs met in council at Bori village, the tribal capital of Ogoniland, on 26 August 1990. Following extensive deliberations, at which Ken Saro-Wiwa was a principal speaker, an 'Ogoni Bill of Rights' was promulgated, which raised still further the ire of the country's military rulers in Lagos and alarm among Shell executives at the company's joint headquarters in The Hague and in London. Indeed, little room was left for any contrary definition, given that its demands included executive autonomy for Ogoniland, an exercising of direct political power, an automatic right of control over all natural resources and the immediate implementation of measures designed to save the people and the land from yet further environmental degradation.

Shell's very worst fears did not take long to materialise. On 30 October, a public demonstration took place against the company, the first of its kind the Delta had ever witnessed. Shell, which had been tipped off beforehand about the plans of the protestors, appealed to the Rivers State Commissioner of Police for an anti-riot squad to afford Shell installations protection. It was an appeal which led to an immediate response, with the commissioner dispatching his mobile police unit, which had the somewhat unpromising and menacing appellation, for the protestors at least, 'Kill and Go'. The unit did just that, shooting dead 80 people and damaging a large number of homes. It was a bloody incident which attracted the appalled attention of the outside world, with the London headquarters of Amnesty International condemning the police action in particularly ringing tones.

In the face of such uncomfortable international attention, the country's governing military council established a commission of inquiry into the killings, which delivered a verdict that the Rivers State Police Mobile Unit had displayed 'a reckless disregard for lives and property'. But the damage was well and truly done, and, as shocked as Saro-Wiwa was at the horrendous loss of life, it did demonstrate the crucially important role that external organisations, such as Amnesty International, could play in the international arena in highlighting the Ogoni cause, a course of action of particular significance given that, at home, avenues of publicity, such as his column in the *Sunday Times*, were being denied him.

Indeed, under official pressure, Saro-Wiwa was also to be denied access to the country's airwaves to get his message across. In short, the pressure was on and he was subsequently obliged to court the assistance of external influences in order for his campaign to make any substantial headway against a military elite which had become exceedingly rich from the country's oil revenues, and an international commercial giant, Shell, which he saw as being not just complicit in the corruption being practised by the officer corps in the Nigerian Army but totally in cahoots with it.

He lost little time in taking the initiative and, as he records in *A Month and a Day*, promptly wrote an addendum to the Ogoni Bill of Rights: 'Without the intervention of the international community, the government of the Federal Republic of Nigeria and the ethnic majority will continue these noxious policies until the Ogoni people are obliterated from the face of the earth.'

Having committed himself to carrying the fight beyond the shores of his native land, Saro-Wiwa initiated a formal relationship with the environmental campaigning organisation Greenpeace, held talks with Amnesty International's International Secretariat in London, travelled on to Switzerland, where he held meetings at the lakeside city of Geneva with UNPO, the Unrepresented Nations and Peoples Organisation, and pleaded his case to the Working Group on Indigenous Populations. But, without doubt, his single biggest achievement was to recruit the interest of the Western press, radio and television. In the autumn of 1992, this bore the fruit of quality publicity with the transmission of Channel 4's television documentary *The Heat of the Moment*, which, more than any other development, gave notice that the plight of the Ogoni, caused primarily by the devastation of so much of their land by the oil companies, and in particular Shell, had moved onto the international agenda.

The year 1992 was particularly significant for the campaign which saw, in December, Saro-Wiwa issuing a written statement giving the oil companies 30 days in which to make substantial reparations for the environmental damage they had caused in Ogoniland. The statement called for early payments to be made to the tune of $4 billion in recompense for the ecological disaster confronting the Ogoni and an additional $6 billion by way of tax and royalties on the oil extracted from their ancestral lands. While it was a statement that received extensive publicity both at home and abroad, it was ignored by both the oil companies and the country's military rulers.

But Saro-Wiwa's initiatives, his demonstrations of courage undaunted, had begun to attract the active sympathy of the more conservative elements within the Ogoni establishment, several of whom had, in the early days of Saro-Wiwa's political evangelism against the state and Shell, expressed, at best, apprehension and, at worst, mistrust.

One such important recruit to the cause was the senior member of the Ogoni establishment Dr Garrick Letton, who, casting his former reservations aside, spoke at MOSOP's 'Ogoni Day' public meetings, which were attracting audiences of hundreds of thousands. Addressing a rally in January 1993, Letton cast caution to the winds and delivered a stinging indictment of what had befallen his people:

> We have woken up to find our lands devastated by those agents of death called oil companies. Our atmosphere has been totally polluted, our lands degraded, our waters contaminated, our trees poisoned, so much so that our flora and fauna have virtually disappeared. We are asking for the restoration of our environment; we are asking for the basic necessities of life: water, electricity, education. But above all we are asking for the right to self-determination so that we can be responsible for our resources and our environment.

Following a token occupation of a Shell site by young Ogonis, Saro-Wiwa used the occasion to declare the company *persona non grata* throughout Ogoni tribal territory. Such a flamboyant declaration, in tandem with the occupation of company property, did, however, have a negative effect among many senior members of the Ogoni establishment, particularly the chiefs, who, given their conservative inclinations, were wary that the mass support now being recruited to Saro-Wiwa's banner could only spell serious trouble.

The signs of a split in the movement now began to appear. Indeed, within days of the January 1993 Ogoni Day, six of the chiefs, all of whom had signed Saro-Wiwa's Ogoni Bill of Rights, sent a message of trust and goodwill to the military governor of Rivers State and, of even greater significance, to Shell as well. Much to the dismay of Saro-Wiwa and his followers, the chiefs gave an assurance to both the governor and Shell that no further demonstrations would take place.

It was, however, in terms of preventing further politicisation of the problem, too little too late. Journalists and television crews continued to arrive in the Delta and were routinely guided by Saro-Wiwa and his rapidly growing band of followers to examples of oil pollution. Visiting environmentalists also toured the region's oilfields and produced reports which illustrated examples of extensive pollution caused by the activities of the oil companies, pollution which, it was claimed, had resulted in the virtual destruction of much of the Ogoni's traditional means of livelihood: farming and fishing. And, like their counterparts at Port Arthur in far-off Texas, local people told their visitors that the 24-hour practice of flaring was poisoning the very air they breathed. They also showed examples of leaking oil pipelines, with oil in many instances pouring out over the ground and draining into the region's creeks and rivers, killing all marine life in its polluting wake.

Such international exposure rang bells of alarm within the ranks of the country's military government and down through the long, secretive corridors of power at Shell. The company now knew only too well that the situation was rapidly deteriorating and that something, clearly, had to be done. True to the spirit and letter of one of its founding fathers, Henri Deterding, 'Mine is a personality which does not readily submerge itself', Shell was not slow to recommend a course of action which, it was suggested, the company itself would principally initiate. It suggested to the country's military junta that it would embark upon a secret surveillance of Ken Saro-Wiwa, his principal followers and the MOSOP organisation. It was an offer the military rulers in Lagos were most unlikely to refuse and, just as the men at Shell had assumed, they did not.

But then, and just as the chiefs of the Ogoni had long feared, violence erupted. On 30 April 1993, as a land clearance project got underway to prepare for the construction of a new pipeline, a clearance which, with appalling insensitivity by Shell, included the gardens of the people of

Biara village, the workforce which was under contract to the company were faced with a highly charged crowd of demonstrators.

An immediate decision was taken to call on the military to restore law and order. In the ensuing confrontation, which lasted for three whole days, the locals were subdued but at a price. One man was shot dead by the security forces, while eleven sustained serious injury. The incident brought a new and deeply troubling dimension to the whole question of the presence of the oil companies in the Delta and, as they had done earlier, the chiefs again made overtures of peace, designed to placate both the government and Shell, and, being particularly mindful that their earlier message had promised that no further demonstrations would take place, asked that specific measures be taken to subdue MOSOP and its followers.

It was a request which, unknown to the chiefs at the time, dovetailed perfectly with Shell's secret surveillance of Ken Saro-Wiwa and his movement, an intelligence-gathering operation the results of which were then passed on to the country's military rulers. Now, here were the traditional rulers of the Ogoni telling that very same military elite of their 'anger and complete disapproval of the lawless activities of certain elements in Ogoni who claim to be operating under MOSOP'. According to the newspaper *Nigerian Tide*, the chiefs went on, in their written statement, to give an assurance to the authorities that they would support 'any action by [the] government to protect life and property of innocent civilians'.

The Biara incident served to illustrate, in particularly stark relief, the gulf now so rapidly opening up between the chiefs, the traditional 'fathers of the people', and the radicals, many of whom were youths regarded by senior members of the Ogoni establishment as being hotheads with an appetite and aptitude for unleashing violence. And the chiefs' old suspicions of intellectuals, as personified by Saro-Wiwa himself and his immediate followers, now surfaced anew.

The fears of the chiefs were to be most promptly realised when, in the wake of their statement to the country's military rulers, in effect requesting that they subdue MOSOP, some of their homes were attacked by rampaging youths. It was a development which filled many among this group of senior citizens, in every sense of the term, with fear and anger, and several of them fled to the urban safety of Port Harcourt.

But the consequences of the incident at Biara village then took a most serious turn, the effects of which are now comprehensively regarded as constituting the actual beginning of the tragedy which was to unfold and eventually cost Ken Saro-Wiwa his life at the end of a hangman's rope.

In his absence – he was travelling in Europe drumming up yet further support for the cause – MOSOP's executive council entered into negotiations with both the government and Shell, negotiations designed to secure financial compensation for the family of the slain villager, the wounded and for the damage inflicted upon homes and gardens during the violence. In the event, Shell promised that it would pay a lump sum of one million naira (approximately £4,000) to the families of the dead man and those whose members had been injured, and that consideration would be given to further payments to individuals whose land would be crossed by the company pipeline. Such an offer, however, was dependent upon permission being given by the villagers for the pipeline's construction.

Alerted to the negotiations, Saro-Wiwa sent an urgent message home that no agreement should be signed until an assessment had been made about the effect the pipeline would have on the local environment. By the time the message reached the negotiating team, however, the talks had concluded, with MOSOP members having returned from Port Harcourt to Biara. And, for the unity of MOSOP, there was worse to come. When the details of the proposed agreement were reported to the people of Biara, it met with considerable hostility, with many protesting that the sum of one million naira was far from adequate to compensate for the death, injury and destruction that had been inflicted upon the community. But the MOSOP team was ill prepared to seek a return to the negotiating table, believing, with some degree of justification, that both the government and Shell would regard such a development as a sign of weakness and further evidence that the unity of the movement was disintegrating.

Indeed, such fears were to be promptly realised when the government, seizing the opportunity to press home an advantage and well armed with a flow of political intelligence being made available to it by Shell, increased its surveillance of Saro-Wiwa, placing a corresponding degree of personal pressure upon him, an aspect of which was his detention, often for several hours, as he passed through Lagos Airport on his way to overseas engagements. One such incident occurred at his attempted departure for Vienna, where he was scheduled to address a meeting of the United Nations Human Rights Conference. On this occasion his passport was seized.

This particular incident took place in June of 1993, on the eve of presidential elections in which the country's military rulers were to make a bid for continuing in power via the ballot box in the hope of gaining both national and international respectability. In the event, it was decided that preventing Saro-Wiwa from attending a UN conference was, perhaps, not the best way to achieve such an objective and he was permitted to proceed on his way, without any charges being laid against him. But, nonetheless, it was a clear indication of what all too soon was to come to pass, and further evidence, if indeed such was needed, of the determination of the government and, indeed, Shell to clip Saro-Wiwa's wings.

It was the occasion of the presidential elections which was to lead not only to Saro-Wiwa's arrest but also to a terminal split within MOSOP. The executive had resolved well before the elections that it would not participate, a resolution that Saro-Wiwa had proposed but which had faced stiff opposition from within the movement. In the event, however, he did carry the vote and on the day of the election, 12 June, a boycott of the polls within Ogoniland was virtually complete – but at an eventual heavy price to MOSOP and its founder.

Yet again it was the hotheaded youths, just as many of the movement's members and the chiefs had feared, who made the running, mounting roadblocks on the way to the polling stations and threatening physical violence against those who expressed their intention to cast their votes. And again, it was the traditional leaders of the Ogoni, the chiefs, who faced intimidating threats from the more radical elements, particularly its youth wing, for their many endeavours designed to encourage people to participate in the elections. These, according to Saro-Wiwa and his supporters, included the practice of 'bussing in' would-be electors, which was seen as a direct challenge to MOSOP and a provocative one at that.

It made for a most unstable situation, one with the promise of yet further violence and widespread unrest. While the military rulers were, themselves, to take the most radical course of action, that is to cancel the elections nationwide, halfway through the actual process, followed by the decision to arrest Ken Saro-Wiwa some ten days after the polls had closed. It was a decision that ignited public protests throughout Ogoniland and split MOSOP into two opposing factions.

In a rapidly deteriorating situation, which posed a most definite threat to public order and was characterised by claim and counter-claim within a now permanently divided movement, including the accusation by Saro-Wiwa and his followers that those who had supported participation in the election were in the substantial pay of Shell, the country's military rulers ordered security forces into Bori, the home of MOSOP's founder and the very heartland of his support.

Ken Saro-Wiwa's world began to fall apart. The movement to which he had devoted so much effort and so much of his life disintegrated, with resignations by some of MOSOP's most senior figures, some of whom claimed that Saro-Wiwa himself had to bear responsibility for the violent elements within the movement, particularly the acts of physical intimidation by its youth wing. The resignations signalled the beginning of a power struggle within MOSOP, but, yet again, there was worse still to come.

Just two months after Saro-Wiwa's arrest, his worst fears that the situation of the despoliation of so much of Ogoniland caused by the extraction of crude oil and the accompanying sense of alienation that such activities by Shell had caused among his people would lead to violence in the Delta, fears that he had so eloquently expressed in Lagos at the reception to launch *On a Darkling Plain*, came to pass.

Even now, some 11 years after the event, it remains unknown who the assailants were, although there have been creditable claims that they arrived in Ogoniland, on the Andoni River, by amphibious craft, the means of transport used by the security forces during their attack on Saro-Wiwa's home village of Bori just two months earlier. On this occasion, however, the assailants' target was the riverside village of Kaa, where they inflicted extensive death and destruction.

It was not the first time that the Ogoni had been attacked by assailants unknown, a similar attack having occurred just a month earlier, in July, during which an estimated 100 Ogonis were shot dead in their boats on the Andoni River. The fact that in both incidents the attackers were armed with modern weapons, including mortars and hand grenades, gives more credence to the theory that they were from the country's security forces. The military government attributed both attacks to tribal conflict, an opinion vehemently refuted by the Ogoni themselves and other tribes in the area.

But as conflict continued into 1994, with the government routinely laying the blame at the Ogonis' door and using such an argument to justify yet further police and military incursions into the tribe's territory, it became increasingly clear that a policy decision had been taken at the most senior of levels within the military establishment not only to foment tribal conflict, which would then be blamed on the Ogoni, but also to exercise a corresponding force in a bid to bring them into line. A clear indication of this was that before the security forces commenced their programme of armed incursions into Ogoni territory, all Ogonis in their number had been reassigned to other duties, so as not to run the risk of having men with divided loyalties at the time of action. It was a policy that had Ken Saro-Wiwa and MOSOP clearly in its sights.

Almost certainly, Shell, whose operations in Ogoniland were now being severely hampered by the rapidly deteriorating security situation, had told the government that order had to be restored before it could continue its work unimpeded, a statement of fact given particular emphasis by the subsequent decision by the company to cease operations until law and order were restored. The authorities were very concerned about this development, as an official instruction of 12 May 1994, to the military governor of Rivers State, Lieutenant-Colonel Dauda Musa Komo, made abundantly plain. It read, in part, 'Shell operations still impossible unless ruthless military operations are undertaken for smooth economic activities.' It also called for surveillance to be increased on 'Ogoni leaders', outlawed visits from foreign human-rights and environmental campaigners to the region and, with regard to the financing of such activities, instructed that 'pressure on oil companies for prompt, regular inputs as discussed' be maintained. It was a funding exercise that, initially, Shell denied most vigorously but later went on to confirm.

The end could not, now, be long in coming and it arrived all too soon for Ken Saro-Wiwa. Just days after the instruction to Lt Col Komo to, in very real effect, create a climate which would make MOSOP's campaigning activities quite untenable, Saro-Wiwa, who had earlier been released from police custody, was travelling to a rally in Ogoniland when his car was stopped by the security forces. He was ordered to proceed no further, to return home, which he agreed to do. What he did not know, however, was that just a short distance from where the rally was to be held, another meeting had been arranged, to be attended by senior establishment figures, men who had long been opposed to the works of the radicals, the intellectuals as personified by Saro-Wiwa. The group included an old political foe, Edward Kobani, his brother Mohammed, Chief Samuel Orage, a one-time Rivers State Commissioner, and his brother, Chief Theophilus Orage, and another conservative activist, Chief Albert Badey, who had also long opposed the radical element. News of their meeting had become known, however, to local youngsters from MOSOP's youth wing, who went to the place where the meeting of elders was taking place and demanded that they show themselves and face the public's wrath. The precise sequence of events is most difficult to establish but violent attacks on four of the men led to their deaths.

The following day, 22 May 1994, Ken Saro-Wiwa was again arrested. Indeed, at a press conference on that very same day, as reported by the American organisation Human Rights Watch, Lt Col Komo made it very clear whom he blamed for the deaths of the elders, commenting that he had given orders for all those associated with the attack '[to] be rounded up. The MOSOP leadership that was part of this game must be arrested.'

The country's military rulers had long been waiting for such a development and they responded in a particularly brutal and unrestrained manner. During the two-month period which followed the killings of the four elders, some sixty villages were attacked, leaving an estimated fifty people dead. Human Rights Watch reported the attacks in full, bloody detail:

> Troops entered towns and villages shooting at random as villagers fled to the surrounding bush. Soldiers and mobile police stormed houses, breaking down doors and windows with their boots, the butts of their guns and machetes. Villagers who crossed their path, including children and the elderly, were severely beaten, forced to pay 'settlement fees' (bribes) and sometimes shot. Many women were raped. Before leaving, troops looted money, food.

Eight long months were to pass before Saro-Wiwa was formally charged, during which time he was denied permission to meet with his legal advisers. But there was worse yet to come.

The country's military government had decreed that the Ogoni leader would not go to trial before a civilian court but, instead, was to be subject to a special tribunal, one with authority to hand down a sentence of death and to which there would be no automatic right of appeal. Saro-Wiwa was finally charged on 28 January 1995, and then, and only then, was he permitted to consult with his lawyers, meetings at which Colonel Paul Okuntimo, the Commander of Rivers State Internal Security Task Force, a man with a reputation for brutality and intimidation, was all too often present.

Both Saro-Wiwa and other senior members of MOSOP who had been arrested at the same time claimed that during their imprisonment they were subjected to routine beatings by their guards, were chained for much of the time, not given adequate rations and refused medical treatment. The conduct of the tribunal, composed of two judges and an army officer, was so irregular and displayed such scant regard for judicial processes that it appalled Saro-Wiwa's defence team. The noted human-rights lawyer Chief Gani Fawehinmi, head of the defence team, was so disgusted that six months into the process he resigned in protest. Indeed, several of the witnesses for the prosecution were later to make sworn affidavits that they had been bribed by officials to bear false witness against Saro-Wiwa and, in fact, no credible evidence was brought before the tribunal to prove that Saro-Wiwa had been complicit in the murder of his opposing Ogoni leaders.

In its account of the trial, *Ogoni: The Struggle Continues*, the World Council of Churches recorded for posterity Saro-Wiwa's closing words to the tribunal. 'I have no doubt at all about the ultimate success of my cause, no matter the trials and tribulations which I and those who believe with me may encounter on our journey. Nor imprisonment nor death can stop our ultimate victory,' a belief supported by the claim expressed throughout his trial that he was entirely innocent of the charges brought against him.

The manner in which the tribunal was conducted attracted condemnation on an international scale. Writing in June of 1995, the distinguished London-based criminal lawyer Michael Birnbaum QC delivered his verdict on the workings of the tribunal. 'It is my view that the breaches of fundamental rights I have identified are so serious as to arouse grave concern that any trial before this tribunal will be fundamentally flawed and unfair.'

When, on 31 October, the tribunal delivered a guilty verdict and sentenced Ken Saro-Wiwa to death by hanging, Birnbaum was no less forthright:

> The judgement of the tribunal is not merely wrong, illogical or perverse. It is downright dishonest. The tribunal consistently advanced arguments which no experienced lawyer could possibly believe to be logical or just. I believe that the tribunal first decided its verdict and then sought for arguments to justify them. No barrel was too deep to be scraped.

Others, however, were not so sure. MOSOP's one-time president, Dr Garrick Letton, told the tribunal:

> Saro-Wiwa must be exposed for what he is. A habitual liar, a person who uses the travails of his own people to achieve his selfish desires and ambitions. A person who is prepared to engineer the elimination of his elders. A person who in this situation cannot escape complicity in the murder of the four prominent Ogoni leaders.

On 8 November, Nigeria's military rulers, through their Provisional Council, confirmed that the death sentence would be carried out. That much-extolled body, 'the international community', raised its voice in protest. Sanctions were threatened should the execution take place but all to no avail. On the morning of 10 November 1995, Ken Saro-Wiwa was taken in chains, together with eight other members of his movement, to Port Harcourt jail, where they were hanged by the neck until death overtook them.

The executions sparked worldwide protests. In far-off Auckland, New Zealand, the Commonwealth Conference was sitting on the very day of the execution. The organisation immediately suspended Nigeria from membership. Governments in the Western world, where the demonstrations of anger were most intense, imposed a range of sanctions against Nigeria's military administration, including a ban on the future sale of arms and travel restrictions upon the junta's principal figures. But the sanction that would have inflicted the maximum damage on the regime, an oil embargo, was not, of course, part of the sanctions package.

On the ground, throughout Ogoni territory, the executions of Ken Saro-Wiwa and his comrades sparked yet further violence and, again, and just as Saro-Wiwa had warned, the Delta experienced a new round of violence bordering on civil war as a wave of unrest, this time amongst the Ijaw people, the largest tribe in the region, swept across the land with the speed and ferocity of an unquenchable bushfire.

While Shell had struck the public attitude widely practised by companies operating in the Third World at times of national crisis – that of maintaining that the arrest, trial and subsequent judicial murder of Ken Saro-Wiwa was an internal matter – the company was visibly shocked and, indeed, deeply troubled by the unrest which erupted in the wake of the hangings, with one of its own reports recording that acts of violence in the Delta more than doubled in the two years following 1995. But the company's silence only served to antagonise its opponents: its critics laid the responsibility for the widespread pollution and environmental devastation and the fact that the people of the Delta were not receiving the revenue from the sale of oil which they resolutely believed was rightly theirs at Shell's door. Indeed, such voices grew louder, more persistent in the wake of Ken Saro-Wiwa's execution.

It was, clearly, a situation to which Shell had to respond. And respond it did, with the principal burden being that it was the government's duty to ensure that revenue from the extraction and sale of crude oil was channelled to those communities in whose territory the company was working. With regard to the leaking of oil from its pipelines, Shell stated that very great technical care had been taken during the construction process to prevent such leaks coming to pass, but when they did, all efforts were promptly made to put the matter to rights and went on to accuse its critics of greatly exaggerating the problem. Then the company went further still.

On 14 November 1995, just four days after Saro-Wiwa was hanged, Shell's managing director in Nigeria, Brian Anderson, fired a company broadside in an internationally distributed press release, which read, in part:

> We totally reject accusations of devastating Ogoniland or the Niger Delta. This has been dramatised out of all proportion. The total land we have acquired for operations to build our facilities, flow-lines, pipelines and roads comes to just 0.3 per cent of the

> Niger Delta. In Ogoniland we have acquired just 0.7 per cent of the land area. These are very small figures that put the scale of our Niger Delta operations firmly in perspective.

Believing, manifestly, that the very best form of defence is attack, such a line was again taken in 1998, with one of Shell's 'Environmental Briefs' issued in London in effect laying the blame for what had come to pass in the Delta firmly at the door of local people. The brief cited the region's 'rapidly expanding population, over farming, deforestation and industry' and, while claiming that the company was always prepared to clean up spills from its pipelines, it accused local people of being primarily responsible for the pollution of the land caused by the leaking crude.

Later that year, the company's London headquarters issued a further brief, 'Oil Spills', which returned to its position that pollution in the Niger Delta generally and in Ogoniland in particular was the direct result of vandalism by local people:

> Sabotage remains a significant problem, despite the widespread awareness that no compensation is paid in such cases. The usual motive for sabotage is to press claims for large sums of money as compensation and/or to attract temporary employment in the subsequent clean-up exercise.

It was a line of defence which, while predictable, did hold some truths – vandalism had indeed long been a problem, for example. Yet the statement chose to ignore the fact that, at the official level, the warnings of the social and political effects of pollution being caused by the oil companies in the Delta had first been issued 12 years earlier.

Indeed, as the Amsterdam branch of Greenpeace International published in one of its 1994 reports, 'Shell Shocked: The Environmental and Social Costs of Living with Shell in Nigeria', the Nigerian National Petroleum Corporation (NNPC), which, in partnership with Shell, operated the extraction of crude oil in Ogoniland, had voiced its concerns in 1983. The NNPC's inspectorate wrote in a report:

> Since the inception of the oil industry in Nigeria more than 25 years ago, there has been no concerned and effective effort on the

> part of the government, let alone the oil operators, to control the environmental problems associated with the industry. We have witnessed the slow poisoning of the waters of this country and the destruction of vegetation and agricultural land by oil spills which occur during petroleum operations.

The writing has, indeed, long been on the proverbial wall and on the oil-stained, poisoned waters of Nigeria's Delta region, a message which gave adequate warning of a coming catastrophe but which was ignored by an overconfident and insensitive international company and a government whose power came all too often from the barrel of a gun as opposed to the people's ballot. It was an unheeded warning, for which, in the tragic event, Ken Saro-Wiwa, for all his political imperfections, both real and imagined in the minds of his opponents, paid with his life. The image, which went around the world, of this man of letters swinging at the end of a hangman's rope inflicted an indelible stain on the reputation of Nigeria and that of Shell.

In his brilliant 2000 work, *This House Has Fallen*, the journalist Karl Maier writes of visiting a member of Shell's public relations team in Nigeria. The venue of the meeting was the Shell Club in Port Harcourt. It is set amidst verdant, well-manicured lawns and serves as a recreational facility for company employees and their families. The club has an excellent range of bars, restaurants, tennis courts, a soccer pitch, satellite television, all of which provide those who experience its pleasures (among whom are, on occasions, members of Nigeria's ruling class) a taste of home, where the squalor and corruption of Nigeria is banished from view. Shell's PR officer told Maier, 'Shell spends a lot of money to build all this, so that for a while we can forget the rest of Nigeria is out there. But of course Nigeria is always there.' To which he could so well, and so justly, have added, as is the shadow of Shell and the darker shadow of a man swinging at the end of a hangman's rope, a disturbing fact which has not inhibited the company's continuing presence in the country.

But perhaps the saddest and most salutary judgement, post Saro-Wiwa, of the industry and Shell's leading role in it was provided by Nigeria's Finance Minister Nedadi Usman. For in May 2004, she was quoted in the magazine *New Internationalist* as saying: 'If we hadn't discovered oil, we would have been better off today.'

CHAPTER 11

THE BEACH AND BEYOND

In 2004, though the issue of ecological destruction caused by the activities of the oil giants in Africa remains unresolved and highly inflammatory, attention has been somewhat deflected by another Shell enterprise in a different corner of the globe which is causing alarm bells to ring with increasing stridency throughout the internationally represented environmental campaigning family.

The danger which the extraction and transportation of oil can pose to the world's oceans has long been a cause of concern and was highlighted back in 1978, the year in which the oil tanker *Amoco Cadiz*, carrying over a million barrels of Shell oil, spilt its cargo on the formerly pristine beaches of western France. On 16 March that year, the 240,000-ton *Amoco Cadiz*, carrying some 223,000 tons of crude oil, was blown onto rocks following the failure of her steering gear. The following day, she broke her back and a quarter of her cargo, some 80,000 tons, went into the sea, coating 100 miles of coastline in thick black crude. Yet there was worse still to come. Eleven days later, the ship's forepeak broke off and the remainder of her cargo poured into the sea, a total of more than 1.5 million barrels. It was the worst disaster of its kind in maritime history.

The consequences were both swift and severe. The company suffered serious sales losses in France and even greater damage to its reputation and image. Added to this were years of litigation for the owners of the stricken vessel and its operators, Amoco. But the cargo was Shell's and the company received savage criticism and comprehensive hostility in

France and, indeed, beyond for failing to have instituted greater security checks on the vessels used to carry its crude oil.

In France, Shell offices were subjected to attack during violent public protests. But what was most instructive throughout the whole tragic, messy affair was Shell's plaintive reaction. 'It's all so very unfair. It was not our fault and, anyway, the five other supertankers at sea that day carrying our cargo were all perfectly safe.'

Plaintive it may have been but the *Amoco Cadiz* affair served notice on Shell that it was, from here on in, very much under public scrutiny. Yet little actually changed within the company and in the strategies it pursued.

Today, the question of marine life and the protection it stands in urgent need of from a current Shell venture has again risen to the top of many environmentalists' agendas. Sakhalin Island is on the far reaches of the Russian Federation, on its far-flung Pacific seaboard, and has long been comprehensively regarded as being of very special environmental significance. For the waters of Sakhalin are the home and breeding ground of one of the planet's rarest and, now, most endangered marine species, the western grey whale. Indeed, it is so very rare that, in the 1970s, when there had not been sightings for several years, many believed the western grey to be extinct.

But Sakhalin is also home to huge reserves of hydrocarbons, in the form of both oil and natural gas, with industry analysts believing that Sakhalin holds up to one billion barrels of oil and enough gas in the island's fields to meet the total global demand for four years, a tempting point not lost on the men at Shell, which, together with its junior partners, Japan's Mitsui and Mitsubishi corporations, established the Sakhalin Energy Company in 1994.

Shell, as the senior partner and driving force behind the enterprise, which has 'offshore' status by being registered in the British Crown Colony of Bermuda in the mid-Atlantic, also signed what is described as 'a sharing agreement' with Moscow, which analysts have established leaves precious little of economic benefit to Russia and, significantly, places severe limitations on the federation's environmental laws. In precise terms, Shell, on this ecologically and environmentally fragile island, is, with a shameless amorality, exercising at least two of its very own styled imperatives, 'capital discipline' – that is evading any working arrangement that will diminish the bottom-line benefit – in tandem with the company's very own brand of 'personal accountability', which translates in its Sakhalin operations, as, indeed, it has elsewhere, into a well-crafted arrangement of words that imply that the company will operate in an environmentally responsible manner but which, in actual fact, is an illusion.

This being manifestly so, marine biologists around the world are alarmed at the presence of Shell on the island, particularly with regard to the damage being done to the western grey whale. In the wake of extensive research, it is now believed that there are only 100 greys left in the Sakhalin feeding grounds and of these only 23 females of reproductive age have been identified. Since 1999, an increasing number of 'skinny whales', emaciated and clearly undernourished, have been observed in the schools off the island.

The crucial question is: are the whales being adversely affected by the development of an oil industry on Sakhalin? That Shell is responsible for the rapidly deteriorating conditions of the whales, allied to a marked reduction in salmon stocks from which one-third of the local population derives its livelihood, David Gordon, director of the Pacific Environment campaigning body, has little doubt, citing the damage to marine life caused by the company having installed platforms off the shores of Sakhalin, which caused sediments to be disturbed and then drift into the grey whales' feeding grounds.

Russia contains one-third of the world's remaining wild salmon stocks and Shell's laying of a pipeline has disturbed, and in some cases fatally so, 59 wild salmon stream beds and waterways vital to the health and continued existence of the fish. Gordon is also convinced that Shell's practice, commencing in 1997, of dumping drilling mud and cuttings – toxic materials that come to the surface during drilling operations – straight into the sea has had a deadly impact on the benthic organisms upon which the whales feed. So far, scientists have been unable to prove beyond doubt that the establishment of an oil industry on Sakhalin is responsible for the decline in the welfare of the western greys but the reason for the absence of such proof, according to conservationists, is that Shell has been interfering with scientific findings.

When oil was first discovered on Sakhalin, the Russian government decreed that Shell, together with the American oil giant Exxon, should meet the costs of research into what effect the development of their industry would have on the whales. A study group of distinguished marine biologists was subsequently appointed. But, says David Gordon, the biologists fell from the oil companies' favour when their findings were critical of the adverse effect that the drilling for oil would have on the region's marine life:

> They started to come up with results that Shell and Exxon did not much like. This led to pressure on the scientists to change their conclusions, to rewrite the language in their reports and I must speak highly of this group of Russian and Americans because they held their ground and said they would only publish good scientific facts.

The result of such an impasse was that Shell and Exxon stopped funding the scientists and turned instead to consulting companies willing to go along with them in arriving at the conclusions they wanted. 'I believe,' says Gordon, 'that the oil companies, Shell and Exxon, are trying to buy up science.'

One of the scientists confirmed to the BBC in the spring of 2004, for the corporation's *File on 4* radio programme, that both Shell and Exxon had made attempts to block research findings, findings which proved beyond reasonable doubt that the western grey whales off Sakhalin Island were, indeed, being adversely affected by the oil installations and badly so.

Nonetheless, Shell's Sakhalin Energy Company went on to issue a statement which declared in robust terms that its activities on Sakhalin and in its waters were not having a negative impact on the whales and that the company 'does not seek to influence research findings'. Sakhalin Energy added that the work of the consultants it had employed had been produced 'independently of Sakhalin Energy'; this had now been published and supported Shell and Exxon's claims that no environmental damage was being done as a result of their activities in the area.

Emboldened by its own rhetoric, based on a report for which it had paid, Sakhalin Energy announced its intention to push ahead with further development on the island, development that is to include the construction of a pipeline that will run the full 800 kilometres of Sakhalin's length and that will carry both crude oil and gas to an export terminal.

It was a statement which caused shock and consternation to

environmentalists around the world, with full and just cause. The whole of the Sakhalin area is notoriously earthquake-prone and in 1995 suffered a major quake which left some 2,000 people dead. Yet despite this, Sakhalin Energy seriously proposes building the pipeline underground, where it will cross 22 known active fault lines. Such a plan assumes an even graver dimension when it is established that at other known seismic hotspots throughout the world, pipelines are usually constructed in cradles above the ground. Such conduct, which is both brazen and indicative of the fact that Shell particularly continues to believe itself invulnerable and, therefore, above public criticism, is yet further evidence of the pivotal importance the company attaches to its Sakhalin operations in the wake of having been found to have grossly inflated its reserves. 'It [Shell] is desperate,' says a senior analyst, 'to find a replacement for the recoverable reserves it fictitiously created.'

In the case of Shell and Exxon's operations on Sakhalin Island, the companies believe that by juggling with scientific evidence they can escape the finger of caution being wagged at them by both individuals and international environmental organisations – yet another example of their appetite for self-delusion. Indeed, so concerned was the World Wildlife Fund (WWF) that, together with the professional and financial support of other non-governmental organisations (NGOs), it commissioned the independent and internationally respected consultant Richard Fineberg, who advised on the construction and siting of the Alaska oil pipeline, one of the largest in the world, to assess the Sakhalin project.

Fineberg was far from being impressed, noting that there was a significant absence of rigour on the safety aspects of the development:

> One of the things I wanted to know was what had been done in terms of risk analysis of the seismic condition at fault crossings. After the question had been put a number of times over several months, I finally received a document indicating an assumption, just an assumption, that the pipeline would withstand an earthquake and that, therefore, there was zero risk. That's not a risk analysis, it's, yet again, merely an assumption. Sakhalin Energy is woefully short of being able to demonstrate that its plans for Sakhalin are safe.

In a statement, the company said that it had received advice from 'foremost experts' but admitted that at the final design stage 'a proper technical review will be performed to determine the safest crossing technique should reviews indicate that the current fault crossing construction plans are not suitable'.

Environmental critics, however, were quick to respond by saying that such a process should, of course, have been done at the very outset. In actual fact, negotiations between environmental and conservation groups and Sakhalin Energy have broken down, and legal action is being taken in the Russian courts to try and force the implementation of the country's strict environmental laws. Moreover, a legal protest over the pipeline is also being mounted. A World Wildlife Fund spokesman confirmed that talks between his organisation and Shell have degenerated into 'an empty dialogue'. 'We have learnt,' he laconically remarks, 'that the fact that a company like Shell is talking to you does not necessarily mean that they are listening to what you have to say.'

It may, indeed, be a dialogue in which Shell, while listening with apparent patience, rarely, if ever, subsequently acts as though it actually heard what had been said, but as Shell's Sakhalin operations moved towards what the company call 'Phase 2' of the project in the summer of 2004, the World Wildlife Fund issued a statement with much of an 'Eleventh Minute of the Eleventh Hour' character about it. It made grim reading but clearly established that there remains much which could still be done to illustrate both the potentially deadly flaws in the project and the means of bringing yet further pressure on Shell to have a change of heart, a process in which several crucially important financial bodies could well play a fundamental role. The statement read, in part:

> Phase 2 of Sakhalin is a US$10 billion oil and gas development. The Sakhalin Energy consortium is led by Shell, with Mitsui and Mitsubishi as key shareholders. The consortium has applied for project finance from the European Bank for Reconstruction and Development, the US Export-Import Bank and the Japanese Bank for International Cooperation. However, the current plans for Phase 2 are fundamentally and seriously flawed on an environmental level.

Under the sub-heading 'Common Demands', the statement continued:

> A coalition of 50 Russian NGOs published a set of common demands in January 2003. WWF supports these demands through its Russian national office, which implements conservation work in the Sea of Okhotsk region surrounding Sakhalin Island.

Continuing with 'One year and no response', the WWF statement then highlighted the unresponsive nature of Shell:

> Shell has now had a full year to respond to the concerns of its stakeholders. The strength of the group is unprecedented. Shell cannot claim that it is not clear what a huge group of stakeholders are asking. NGOs do not believe the project as it stands meets all the environmental requirements of the finance institutions. NGOs are making legal challenges to the project in Russia.

The statement then proceeded to explain the environmental concerns raised by Shell's Phase 2 plans:

> Phase 2 of Sakhalin involves the installation of an offshore platform on an existing oilfield and the installation of a single large platform on a gasfield. These platforms, as well as one other, will be linked to the shore by offshore pipelines. The oil and gas will then be transported via 800 kilometres of onshore pipelines to Prigorodnoye, in the south of Sakhalin Island and the export terminals.

Addressing one of the WWF's principal concerns, the western grey whale, the statement confirmed that:

> The grey whale is on the endangered species lists of the United States and Russia, and has been recognised as being critically endangered by the International Union for the Conservation of Nature (IUCN). Recent scientific evidence suggests that less than 100 individuals and possibly fewer than 20 reproductive

> females capable of bearing calves remain. This trend is enough to warrant reassessment of the project. Shell has offered US$5 million for further whale monitoring and research over five years, but it is no consolation to know that their decline will be monitored. Whilst Shell is willing to consult on the activities of the research programme, it will not reconsider the project's design.

Under sub-headings 'Flaws' and 'Urgent Priorities', the statement then identified specific dangers inherent in Phase 2.

> The project mirrors flaws in other recent huge infrastructure projects, with local legislation undermined and no strategic environmental assessment having been carried out. The onshore pipeline is located in an area of high seismic activity, only 40 kilometres from the site of the 1995 Neftegorsk earthquake. This region can experience events registering 9 on the Richter Scale, which was experienced on Sakhalin in 2000. The pipe will cross over 1,000 rivers, including waterways essential for the spawning of endangered salmon. Dredging of the site proposed for the Liquid Natural Gas (LNG) plant will produce one million tons of spoil which Shell intend to dump in Aniva Bay, an essential area for the local fishing industry. At this stage of the project, WWF sees the four key issues as: the location of the offshore pipeline; the location of the offshore platform; onshore river crossing techniques; LNG plant dredging spoil disposal. The remainder of the NGO demands, held in common, are also vital to ensure protection of biodiversity in the long term.

The statement concluded with a damning indictment of Shell's much-trumpeted 'Profits and Principles' slogan:

> Shell's latest advertising campaign 'Profits and Principles' is based on the premise that it is not necessary to choose between these two elements. In this case, Shell has clearly chosen its profits over the principle of responding to concerns about the impact of its operations on biodiversity. Putting a price on the survival of the endangered grey whale population is not acceptable.

It is a statement that is as much a challenge to Shell as a detailed confirmation of an internationally held list of concerns about the company's Sakhalin operations and particularly Phase 2 of the project. Indeed, the statement is considered to be of such crucial significance that the decision was taken to afford it considerable coverage in this book principally because of the scientific quality of the conclusions on which it is based but also because it gives the lie to Shell's often 'off-the-record' claim that those who campaign on environmental issues are 'the anorak and open-toed sandal-wearing class, the lentil-munching, left-wing crazed, bearded and unwashed tribe who have nothing better to do'.

Greg Muttit, a senior oil analyst, has commented:

> Shell is, in terms both pure and simple, just not delivering on its environmental promises for the very basic reason that those responsible for the company's environmental policy are not the big players. There are people in the company who believe that reform of the organisation is possible, that it is possible for Shell to work in a more ethical manner and there are those in the company who work very diligently to try and make this happen. But most of these people are not permitted to exercise their influence at the heart, the very core, of Shell's business. They exist as a 'bolt-on'. They are a department in their own right and part of the problem is that people who are good in those sort of roles tend not to make good business managers. What's happening in some of the really big problem areas is that the management are keenly aware of Shell's unimpressive financial performance, and I suspect that they are starting to become obsessed with the figures.

There seems to be little evidence of awareness that a company of the size and significance of Shell can no longer conduct its affairs in the style and practice of a nineteenth-century free-market buccaneer whose only interest is making money, virtually at any price, and using any means at its disposal. Indeed, the essential lesson Shell, caring or otherwise, has to learn is that their 'price' is something that the greater majority, right across the face of the earth, are no longer prepared to pay.

In 2004, it became more evident by the day that the company's financial difficulties had started to throw its other problems into relief, a circumstance that has prompted the big cats in the investment jungle to grow and flex their not inconsiderable muscle. Indeed, on the eve of Shell's Annual General Meeting, on 28 June 2004, came reports that City institutions who control SRI shares (Socially Responsible Investment – the so-called 'Ethical Funds') had started to sell their shares in Shell as a direct result of their concern at the manner in which the company deals with environmental issues in its fields of operation, such as in Nigeria, the United States and Sakhalin Island. Blue-chip City institutions made it publicly known that they had dumped all their shares in Shell, for example Investec Henderson Crosthwaite and Moreley Fund Management disposed of all its Shell shares in the wake of the company's confirmation, in January 2004, of the overestimation of its reserves. Such moves were particularly significant given that decisions by SRI funds exercise a strong influence on mainstream analysts' investment recommendations. Pivotal in the decision to sell was the growing belief that the systems of management Shell had in place were failing and investors demanded to be told exactly what the board of Shell intended to do in order to turn the company around in regard to both its financial performance and its governance.

Such problems were compounded halfway through the summer when Shell was faced with evidence of the price that it may well have to pay when confronted with the public's wrath. Indeed, there are some who believe that so great will current legal challenges to the company prove that, in the event, they could pose the single biggest threat to its continued existence and, most certainly, to its continued existence in its present form.

Take as just one example the legal action being mounted in the United States by Melvin Weiss of Milberg, Weiss, Bershad, Hynes and Lerach, the world's biggest class action specialists, based in New York. Weiss is seeking judicial redress for Shell's American stockholders in a multi-billion-dollar lawsuit and is entirely serious. The legal challenge being pursued by the long-suffering people of Port Arthur in the state of Texas is proving to be of no small significance either.

In short, sharp and brutal terms, Shell was now seen to be 'all at sea' by both its investors and the wider public world. The hour of Shell's extremis was at hand.

EPILOGUE

TWO INTO ONE AT LAST

The *Financial Times* had no doubt about the big story of the day. On Friday, 29 October 2004, the paper's front page proclaimed 'Shell to embark on radical overhaul' in a four-deck headline. In addition to a leading article headed 'Shell becomes a normal company', the whole of page twenty-one was devoted to another three stories, a five-column picture and double-column graphic of the oil giant's reorganisation.

The *Daily Mail* – a paper which from the time of Marcus Samuel's chairmanship of Shell Trading and Transport has maintained an informed and far from uncritical interest in the company's affairs – carried four stories and led its finance section with City editor Alex Brummer's commentary headlined 'Disquiet as Shell goes Dutch'.

And that, in a five-word nutshell, was the essence of the matter. For at a conference called the day before, Shell announced that after 97 years it was at last abandoning its Byzantine structure of twin boards and headquarters, several chairmen and more management committees – including a committee of managing directors – than you could shake a stick at.

Now, after decades of huffing, puffing and procrastination, Royal Dutch and Shell Transport and Trading – the company founded by Marcus Samuel in 1897 – will be fully merged into Royal Dutch Shell plc. The unified company will be headquartered in The Hague but will be listed on the London Stock Exchange, where the newly created giant, with a value of about £100 billion, will have immense power on the

FTSE 100. Currently, the Shell Transport and Trading holding company is valued at about £40 billion.

The new arrangements, according to both British and Dutch newspapers, appear to have substantially allayed fears in the UK of a Dutch putsch and of a British coup in Holland. Royal Dutch Shell's chief executive will be Jeroen van der Veer, the current chairman. Aad Jacobs will be the new company's chairman. The committee of managing directors will be jettisoned but its members will find places on the board of the unified company. Current non-executive board members are expected to leave soon and Lord Oxburgh, chairman of Shell Transport and Trading, was quoted as saying 'new blood' for the board was a priority. It is expected that an outside chairman will be appointed in 2006.

Although Shell's London centre at Waterloo on the South Bank is to be retained, its importance has been much diminished with the transfer to The Hague of 200 top posts. About 3,000 people are currently employed by Shell in London; insiders believe that unification will enable the company to cut 400 jobs. The old tradition of the top job alternating between British and Dutch executives has also been dumped.

The downgrading of London was not met with universal rejoicing. Alex Brummer in the *Daily Mail* said the move needed further explanation:

> The present trend is for overseas companies . . . to move headquarters to London, not in the other direction.
>
> London offers the deepest liquidity in its financial markets, the largest foreign exchange operations and access to the top investment banks. The Hague is a wonderful city, but it is not a top-line financial centre.

Moreover, many observers, noting the enthusiasm with which senior Shell people were talking up the new arrangements while heaping scorn on the old and fiendishly complicated corporate set-up, wondered why it had taken the company so long to change if the benefits were as obvious as was now being claimed.

Most City commentators and financial writers said that the principal advantage of the adoption by Shell of a single, sane and vastly simplified structure was that it would give the company better access to capital markets and permit the pursuit of 'all-paper' takeovers. And acquisitions could help Shell deal with its longstanding hunger for crude and its urgent need to find oil in quantity.

For although the company was the same day able – by virtue of oil prices of $50+ per barrel in London for Brent and $55+ in the US for light crude – to report the record doubling of third-quarter profits to $5.4 billion (£2.95 billion), taking the nine-month surplus to a stunning $14 billion, none of Shell's many problems had disappeared.

And one measure of difficulties still to be addressed was the fact that even as the company was generating profits at the incendiary rate of £1.3 million per hour, Shell was also on 29 October obliged to report not only that oil production had fallen by 4 per cent but also that a further 900 million barrels might be slashed from proven reserves already cut four times since January.

The announcement was made by Malcolm Brinded, head of exploration and production. It was, he said, 'disappointing to be addressing a subject that we had put behind us. People may ask why we have made changes to figures that we said were correct before.'

Part of the answer was that Shell had so far only rechecked 8 billion of the 14.35 billion barrels of reserves it said were compliant with the US Securities and Exchange Commission regulations. The possibility of further downward revisions was thus all too clearly left open.

The point was not lost on Paul Durman in the *Sunday Times*. He said that Shell now has a third less oil under the ground than it claimed only 12 months ago and would probably not replace even 60 per cent of what it pumps this year (2004). 'When an oil company can't find oil – and isn't even too sure about the stuff it has found – it's time to start worrying,' he wrote.

In the United States, meanwhile, Shell was still facing the possibility of a massively damaging class action being brought by New York lawyers on behalf of pension fund investors in the wake of the original January reserves fiasco.

Coughing, gasping residents of 'Gasoline Alley', the area of Port Arthur in Texas all too literally within spitting distance of the giant Motiva refinery complex, are continuing their battle with the company to be able to breathe clean, non-toxic air.

In 'Cancer Alley', on the banks of the Mississippi between New

Orleans and the Louisiana state capital Baton Rouge, *The Guardian* reported that the black community of Diamond is disappearing at its own request.

Diamond is where within the memory of people not much more than middle-aged, cat and crayfish were caught, and watermelons, sugarcane and even tangerines were grown. Now the area is home to a huge concentration of refineries and chemical works, including the ethylene- and propylene-producing plant of Shell Chemicals.

After accidents, fatal explosions and a bitter 11-year campaign, the paper said, Shell had finally been shamed into abandoning its policy of buying up neighbouring residential property at prices reflecting the fact that nobody in their right mind wanted to live there. Just three years ago, the company was obliged to agree to purchasing homes at fair prices.

And in Nigeria, the Ijaw people launched claims for $1.5 billion for hurt and damage inflicted by Shell in the Delta region. The Nigerian trade union organisation branded the company an 'enemy of the people'.

In the first days of December 2004, Nigerian environmental activists and protestors found a surprising ally in Washington. For the Senate's Committee on Compliance recommended sanctions against the Shell Petroleum Development Company (SPDC) over its non-compliance with a resolution of both houses of the National Assembly directing the company to pay compensation of $1.5 billion to Ijaw aboriginals in Bayelsa State for oil spillages and pollution that have occured over the years.

Corporate comfort of a kind was, however, provided by the *Financial Times*' leader warmly welcoming Shell's long overdue discovery of the virtues of structural simplicity. But a magisterial summing up contained a sting in the tail, for the paper said: 'Shell's task now is to put the same energy into operations as it has into corporate reform, and redouble its checks on existing oil reserves and its efforts to find new ones.'

On Friday, 4 February – and as this book was well on its way to the printers – Shell unveiled the biggest profits in its history and earnings in 2004 of $18.5 billion (£9.82 billion), or rather more than £1 million per hour.

That, for shareholders and investors, was the good news. The bad news, as encapsulated in the first paragraph of the front-page lead story in the *Financial Times*, was that Royal Dutch/Shell had 'cut its proved oil and gas reserves by another 10 per cent and was still struggling to replace the oil and gas it extracted from the ground'. The latest cut – the fifth Shell has publicly announced – amounted to another 1.4 billion barrels. Cumulatively, the cuts have slashed Shell's reserves to a total now currently estimated at 12.95 billion.

The paper went on to report that Shell had only replaced 15–25 per cent of the depletion of its reserves in 2004, adding that the RRR was expected to recover next year, although still falling well short of the levels achieved by competitors such as ExxonMobil and BP. Investors were also warned to expect further cuts in October.

When he announced the results, Shell chairman Jeroen van der Veer – who in December 2004 said his head would be on the block if he failed to get to grips with the company's problems by the end of this year – said of the reserves crisis: 'We have done everything possible to get this right. It has been a huge effort to get this behind us.'

It sounded all too familiar and, in the face of truly colossal, record-shattering profits, the market delivered its judgement – that you still could not be sure of Shell – by marking the company's shares down 8.25 pence to 471.75 pence in London.

LIST OF ABBREVIATIONS

BNOC	British National Oil Corporation
bpd	barrels per day
CMD	committee of managing directors
E&P	exploration and production
EOR	enhanced oil recovery
EPA	Environmental Protection Agency
Ex-IM	US Export-Import Bank
FSA	Financial Services Authority
IPC	Iraq Petroleum Company
IUCN	International Union for the Conservation of Nature
LNG	Liquid Natural Gas
MOSOP	Movement for the Survival of the Ogoni People
MTD	Maria Theresa Dollar
NGO	non-governmental organisation
NNPC	Nigerian National Petroleum Corporation
NOC	National Oil Corporation
OPEC	Organisation of Petroleum Exporting Countries
PDO	Petroleum Development Oman
PFLOAG	People's Front for the Liberation of the Occupied Arabian Gulf
RRR	ratio of reserve replacement
SBM	single buoy mooring
SEC	Securities and Exchange Commission
SPDC	Shell Petroleum Development Company
SRI	Socially Responsible Investment
TAGA	Trace Atmosphere Gas Analyser
TNT	tri-nitro-toluene
UNPO	Unrepresented Nations and Peoples Organisation
VLCC	very large crude carrier
VNR	Video News Release
WWF	World Wildlife Fund

BIBLIOGRAPHY

Aburish, Said K., *The Rise, Corrpution and Coming Fall of the House of Saud* (Bloomsbury, London, 1994)
Achebe, Chinua, *Anthills of the Savannah* (Penguin, London, 1987)
Beasant, John, *Oman: The True-Life Drama and Intrigue of an Arab State* (Mainstream, Edinburgh, 2002)
Beder, Sharon, *Global Spin* (Green Books, London, 1997)
Benn, Tony, *The Benn Diaries* (Hutchinson, London, 1989)
Biddle, Gordon, and Nock, O.S., *The Railway Heritage of Britain* (Michael Joseph, London, 1983)
Briggs, Asa, *Victorian People* (University of Chicago Press, Chicago, 1975)
Brogan, Hugh, *The Pelican History of the United States of America* (Pelican, London, 1986)
Calder, Angus, *The People's War* (Panther, London, 1971)
Cannon, John (ed.), *The Oxford Companion to British History* (Oxford University Press, Oxford, 1997)
Chekov, Anton, *The Island* (Century Hutchinson, London)
Cooke, Alistair, *Alistair Cooke's America* (Alfred A. Knopf, New York, 1973)
de la Billière, General Sir Peter, *Looking for Trouble* (HarperCollins, London, 1995)
Evans, Harold, *The American Century* (Jonathan Cape-Pimlico, London, 1998)
Fiennes, Ranulph, *Where Soldiers Fear to Tread* (Hodder and Stoughton, London, 1975)

Gilbert, Martin, *The Day the War Ended* (HarperCollins, London, 1993)
Halliday, Fred, *Arabia without Sultans* (Pelican, London, 1979)
Hammer, Armand, *Hammer: Witness to History* (Coronet, London, 1988)
Hennessy, Peter, *Never Again* (Vintage, London, 1993)
Henriques, Robert, *Marcus Samuel: First Viscount Bearsted* (Barrie and Rockliff, London, 1960)
Howarth, Stephen, *A Century in Oil* (Weidenfeld and Nicolson, London, 1997)
Judd, Dennis, *Empire* (Fontana, London, 1997)
Judd, Dennis and Surridge, Keith, *The Boer War* (John Murray, London, 2002)
Keegan, John, *The First World War* (Hutchinson, London, 1998)
Kurlansky, Mark, *Salt* (Jonathan Cape, London, 2002)
Maier, Karl, *This House Has Fallen* (Allen Lane – Penguin, London, 2002)
Mansfield, Peter, *The Arabs* (Pelican, London, 1978)
Monbiot, George, *Captive State* (Pan, London, 2000)
Morris, Jan, *Sultan in Oman* (Arrow, London, 1990)
Pakenham, Thomas, *The Scramble for Africa* (Abacus, London, 1992)
Palmer, Alan, *Dictionary of the British Empire and Commonwealth* (John Murray, London, 1996)
Philbrick, Nathaniel, *In the Heart of the Sea* (HarperCollins, London, 2000)
Phillips, Wendell, *Oman: A History* (Librairie du Liban, Beirut, 1971)
Read, Anthony and Fisher, David, *The Fall of Berlin* (Pimlico, London, 1993)
Roberts, Glyn, *The Most Powerful Man in the Word* (Corvici Friede, New York, 1938)
Roberts, J.M., *Twentieth Century* (Allen Lane – Penguin, London, 1999)
Sampson, Anthony, *The Anatomy of Britain* (Hodder and Stoughton, London, 1965)
Sampson, Anthony, *The Seven Sisters* (Coronet, London, 1988)
Shell, *The Petroleum Handbook* (Elsevier, Amsterdam, 1983)
Skeet, Ian, *Muscat & Oman* (Travel Book Club, London, 1975)
Strachen, Hew, *The First World War* (Simon and Schuster, London, 2003)

Thesiger, Wilfred, *Arabian Sands* (Penguin, London, 1964)
Wheeler, Robert R. and Whited, Maurine, *Oil from Prospect to Pipeline* (Gulf Publishing Company, Houston, 1981)
Winder, Robert, *Bloody Foreigners* (Little Brown, London, 2004)
Yergin, Daniel, *The Prize* (Simon and Schuster, London, 1991)

NEWSPAPERS, MAGAZINES AND INFORMATION SERVICES:

Alexander's Gas and Oil Connections, BBC Radio – *File on 4*, BBC TV – *The Money Programme*, *Channel Four News*, *Daily Mail*, *Daily Telegraph*, *The Economist*, *Financial Times*, *The Guardian*, *The Independent*, *National Geographic*, *New Internationalist*, *Newsweek*, *New York Times*, *The Observer*, *Reuters*, *The Scotsman*, *Sunday Telegraph*, *Sunday Times*, *The Times*, *Wall Street Journal*, *The Week*.

INDEX

INDEX